Studies in Systems, Decision and Control

Volume 204

Series editor

Janusz Kacprzyk, Systems Research Institute, Polish Academy of Sciences, Warsaw, Poland

The series "Studies in Systems, Decision and Control" (SSDC) covers both new developments and advances, as well as the state of the art, in the various areas of broadly perceived systems, decision making and control–quickly, up to date and with a high quality. The intent is to cover the theory, applications, and perspectives on the state of the art and future developments relevant to systems, decision making, control, complex processes and related areas, as embedded in the fields of engineering, computer science, physics, economics, social and life sciences, as well as the paradigms and methodologies behind them. The series contains monographs, textbooks, lecture notes and edited volumes in systems, decision making and control spanning the areas of Cyber-Physical Systems, Autonomous Systems, Sensor Networks, Control Systems, Energy Systems, Automotive Systems, Biological Systems, Vehicular Networking and Connected Vehicles, Aerospace Systems, Automation, Manufacturing, Smart Grids, Nonlinear Systems, Power Systems, Robotics, Social Systems, Economic Systems and other. Of particular value to both the contributors and the readership are the short publication timeframe and the world-wide distribution and exposure which enable both a wide and rapid dissemination of research output.
** Indexing: The books of this series are submitted to ISI, SCOPUS, DBLP, Ulrichs, MathSciNet, Current Mathematical Publications, Mathematical Reviews, Zentralblatt Math: MetaPress and Springerlink.

More information about this series at http://www.springer.com/series/13304

Weichao Sun · Huijun Gao ·
Peng Shi

Advanced Control for Vehicle Active Suspension Systems

Springer

Weichao Sun
School of Astronautics
Harbin Institute of Technology
Harbin, China

Huijun Gao
School of Astronautics
Harbin Institute of Technology
Harbin, China

Peng Shi
School of Electrical and Electronic
Engineering
University of Adelaide
Adelaide, SA, Australia

ISSN 2198-4182 ISSN 2198-4190 (electronic)
Studies in Systems, Decision and Control
ISBN 978-3-030-15784-5 ISBN 978-3-030-15785-2 (eBook)
https://doi.org/10.1007/978-3-030-15785-2

Library of Congress Control Number: 2019934780

This Springer imprint is published by the registered company Springer Nature Switzerland AG
The registered company address is: Gewerbestrasse 11, 6330 Cham, Switzerland

Preface

With increased requirements for vehicle performances, vehicle suspension systems are of importance for contributing to the cars handling and keeping vehicle occupants comfortable and reasonably well isolated from road noise, bumps, vibrations, etc. A well-designed suspension system can effectively promote the whole performances of automobile chassis. Basically, vehicle suspension system consists of wishbone, spring, and shock absorber to transmit and filter all forces between car body and the road. From a perspective of the control mode, vehicle suspensions can be categorized into three types: passive, semi-active, and active suspensions. Among the three kinds of suspensions, active suspensions have the greatest potential to improve the ride comfort and vehicle maneuverability, and this research area has remained attractive for many years.

In active suspensions, actuators are placed between the car body and wheel-axle parallel to the suspension elements, and are able to both add and dissipate energy from the system, which enables the suspension to control the attitude of the vehicle, to reduce the effects of braking and the vehicle roll during cornering maneuvers to increase ride comfort and vehicle road handling. Although active suspensions have many advantages, some problems are also needed to be solved urgently. The main limitative factors of active suspensions fall into three areas: (1) difficult control algorithm; (2) potential risk in reliability; and (3) extra energy consumption.

Focused on the above three aspect problems, the book is organized as eight chapters. Chapter 1 introduces the background, modeling, and problem statements of active suspensions, which can be viewed as the fundamental description of active suspension control. Chapter 2 is concerned with the constrained H_∞ control approaches of active suspension systems in the entire frequency domain, which mainly concentrates on linear convex optimization approach in H_∞ sense. Chapter 3 focuses on the state feedback and dynamic output feedback controller in the finite frequency domain which people are most sensitive to. Chapter 4 aims at nonlinear constrained tracking control via terminal sliding-mode control and adaptive robust theory. Chapter 5 is mainly about the controller design of active suspensions when actuator saturation is taken into consideration. Chapter 6 focuses on the reliability control of active suspension systems, where several kinds of the most possible problems in actuators are considered in controller design. Chapter 7 considers actuator dynamics in the controller design to improve the accuracy, and the electro-hydraulic systems are exampled as actuators to supply the active forces into suspension systems. Chapter 8 carries out active suspension control from an energy point of view, and the energy regeneration scheme and self-powered criterion of motor-driven active suspension systems are investigated.

To summarize, this book presents the most recent theoretical findings on control issues for active suspension systems. By integrating novel ideas, fresh insights, and rigorous results in a systematic way, this book is aimed at providing a base for further theoretical research as well as a design guide for engineering applications of active suspensions. This book can serve as a reference to the main research issues and results on active suspension systems for researchers devoted to control theory or vehicle dynamics control, as well as a material for graduate and undergraduate students interested in control theory and vehicle suspension systems. Some prerequisites for reading this book include linear system theory, matrix theory, mathematics, adaptive control theory, and so on.

Acknowledgements

We are deeply indebted to Prof. Okyay Kaynak (the Electrical and Electronic Engineering Department, Bogazici University), Prof. Hong Chen (Department of Control Science and Engineering, Jilin University), and Prof. Bin Yao (Department of Mechanical Engineering, Purdue University, USA), for their excellent research results which inspire our thinking to finish this book. Weichao Sun would also like to thank his students Shuai Yan, Huihui Pan, and Qian Zhang, who are now working or studying all over the world in various occupations, for their great contributions and detailed discussions.

The financial support of China Automobile Industry Innovation and Development Joint Fund (U1564213) and National Natural Science Foundation of China (No. 61773135, 61790564) are gratefully acknowledged.

Harbin, China
February 2019

Weichao Sun
Huijun Gao
Peng Shi

Contents

Chapter 1
Background, Modelling and Problem Statements of Active Suspensions

In this chapter, the background knowledge of the suspension systems is demonstrated. We mainly introduce three categories of suspension systems: two degree-of-freedom (DOF) quarter-car model, four DOF half-car model, and seven DOF full-vehicle model, whose mathematical models are established. The road excitement model is given in time domain and in the form of spectrum as well. In addition, control objectives of active suspension systems are illustrated.

1.1 Introduction

Suspension systems transmit all forces between the vehicle body and the road and thereby mainly determine ride comfort, road holding and ride safety. The handling capabilities of a vehicle are significantly influenced by the dynamic behavior of the suspension system, i.e., performance improvements of suspension systems can not only make a positive impact on the driver's comfort, but also prevent physical fatigue of the driver and reduce the number of traffic fatalities [1, 2].

Roughly speaking, vehicle suspensions can be grouped into three types: passive, semiactive and active suspensions. Passive suspension systems comprise springs and dampers inserted between the body of vehicle and the wheel-axle assembly. Passive suspensions have the advantages of simple mechanism, easy implementation and high reliability, but they are inadequate in improving ride comfort or road holding for the reason that invariant spring and damper characteristics are unable to cope with different road conditions and conflicting criteria [3, 4].

Semi-active suspension systems feature variable dampers or springs, which means that the damping coefficients or the spring stiffness can be adjusted within a given range. Due to their low energy consumption and high reliability, they are available in a wide range of production vehicles [5, 6]. However, the resulting damper forces

W. Sun et al., *Advanced Control for Vehicle Active Suspension Systems*, Studies in Systems, Decision and Control 204,
https://doi.org/10.1007/978-3-030-15785-2_1

or spring forces are restricted by passivity constraints, i.e., they can only counteract the relative motion of the damper and dissipate energy passively, which are limited in improving ride comfort although they represent a considerable improvement over passive suspension systems.

Compared with the former two kinds of suspensions, active suspension systems require a power supply and are able to generate independent forces of the relative suspension motion. However, due to their energy requirements as well as weight and packaging aspects, active suspension systems have not been integrated in production vehicles, but undoubtedly, active suspensions will be the trend of future vehicle suspension design [7, 8].

The success of design of active suspension systems is determined by two stages. The first stage is to construct a control-oriented dynamic model of vehicle active suspensions; and the second stage is to design and choose a suitable control strategy, which has significant impact on the ride comfort and ride safety. In the following, the classical active suspension models would be presented.

1.2 Control-Oriented Active Suspension Models

Vehicle dynamic modeling is an important step in the design of suspension systems. Generally speaking, the vehicle dynamic model of real vehicle is some degrees of approximation. According to the requirement of controller design, the three dynamic models: two DOF quarter-car model, four DOF half-car model, and seven DOF full-vehicle model, are often used for the theoretical analysis and design of suspension systems. In this section, the quarter-car model is first reviewed, and then half-car and full-car models are discussed.

1.2.1 Quarter-Car Model

If the motions of the four wheels are assumed to be decoupled and the suspension dynamics are only considered in the frequency range of interest for the vertical vehicle dynamics (0–25 Hz), the quarter-car model represents an appropriate modeling framework, which has been used extensively in the literature and captures many important characteristics of more detailed models. It consists of the dynamic behavior of the unsprung mass (representing the mass of a tire, the wheel, the brake, the wheel carrier, and parts of the suspension system) and the sprung mass (mainly determined by a quarter of the chassis mass, including passengers and vehicle payload), connected by the suspension system. Moreover, the tire in this model can be represented by a parallel spring and damper configuration. Figure 1.1 illustrates the mentioned quarter-car model, defined through relation (1.1).

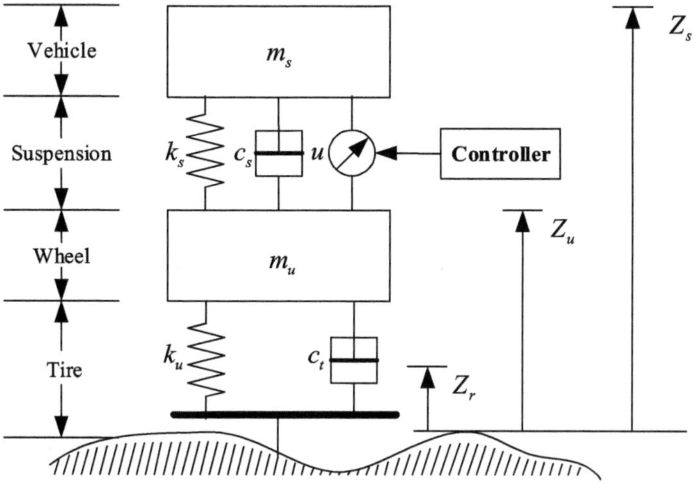

Fig. 1.1 Quarter-Car active suspension model (absolute displacement)

$$\begin{cases} m_s\ddot{Z}_s = -k_s(Z_s - Z_u - \Delta_s) - c_s(\dot{Z}_s - \dot{Z}_t) - m_s g + u \\ m_u\ddot{Z}_u = k_s(Z_s - Z_u - \Delta_s) + c_s(\dot{Z}_s - \dot{Z}_u) - k_u(Z_u - Z_r - \Delta_t) \\ \qquad -c_t(\dot{Z}_u - \dot{Z}_r) - m_u g - u \\ Z_u - Z_r < \Delta_t \end{cases} \tag{1.1}$$

where m_s is the quarter-car body mass; m_u is the unsprung mass (tire, wheel, brake calliper, suspension links, etc.), and g is the gravitational constant. Z_s is the vertical position of the car body, Z_u is the vertical position of the unsprung mass, and Z_r is the vertical position of the road profile. $k_s \in \mathbb{R}^+$ and $k_u \in \mathbb{R}^+$ are the stiffness of the suspension spring and the tire, respectively; $\Delta_s \in \mathbb{R}^+$ and $\Delta_t \in \mathbb{R}^+$ are the length of the unloaded suspension spring and tire, respectively. $c_s \in \mathbb{R}^+$ and $c_t \in \mathbb{R}^+$ are the damping coefficients of the sprung shock-absorber and the tire, respectively. u is the actuator input force. This last inequality of (1.1) is referred to as the passivity-constraint of an active suspension, which is used to guarantee that the actuator only dissipates energy [9]. Based on model (1.1), the system equilibrium point is derived as follows:

$$\begin{cases} -k_s(Z_s^{eq} - Z_u^{eq} - L) - m_s g = 0 \\ k_s(Z_s^{eq} - Z_u^{eq} - L) - k_u(Z_u^{eq} - Z_r^{eq} - R_t) - m_u g = 0 \end{cases} \tag{1.2}$$

where Z_s^{eq}, Z_u^{eq} and Z_r^{eq} represent the corresponding positions in system equilibrium point, and L and R_t represent the length of the suspension spring and tire when the system runs in equilibrium point. Consequently, the solution is given as

$$\begin{bmatrix} Z_s^{eq} \\ Z_u^{eq} \end{bmatrix} = \begin{bmatrix} -k_s & k_s \\ k_s & -k_s - k_u \end{bmatrix}^{-1} \begin{bmatrix} m_s g - k_s L \\ m_u g + k_s L - k_u R_t - k_u Z_r^{eq} \end{bmatrix} \tag{1.3}$$

Then, by choosing $Z_r^{eq} = 0$, the equilibrium point may be rewritten as:

$$\begin{bmatrix} Z_s^{eq} \\ Z_u^{eq} \end{bmatrix} = \begin{bmatrix} L - \frac{m_s g}{k_s} + R_t - \frac{(m_s+m_u)g}{k_u} \\ R_t - \frac{(m_s+m_u)g}{k_u} \end{bmatrix} \tag{1.4}$$

This equilibrium point will then be used to simplify the system model, in order to consider only the dynamical parts.

Around the equilibrium point (1.4), the following dynamical model is thus commonly used:

$$\begin{cases} m_s\ddot{z}_s = -k_s(z_s - z_u) - c_s(\dot{z}_s - \dot{z}_u) + u \\ m_u\ddot{z}_t = k_s(z_s - z_u) + c_s(\dot{z}_s - \dot{z}_u) - k_u(z_u - z_r) - c_t(\dot{z}_u - \dot{z}_r) - u \end{cases} \tag{1.5}$$

where z_s and z_u are the displacements of the sprung and unsprung masses, respectively, z_r is the road displacement input, which is shown in Fig. 1.2. The other variables have the same meaning with (1.1).

Remark 1.1 In (1.1), the damper force and spring force are assumed as linear variation, which results that the quarter-car dynamic model behaves as a linear system. The linear model is popular because of many advantages, such as convenience for performance analysis, easines to be controlled and so on. However, since the actual damping coefficient c_s is always different in the process of the extension and

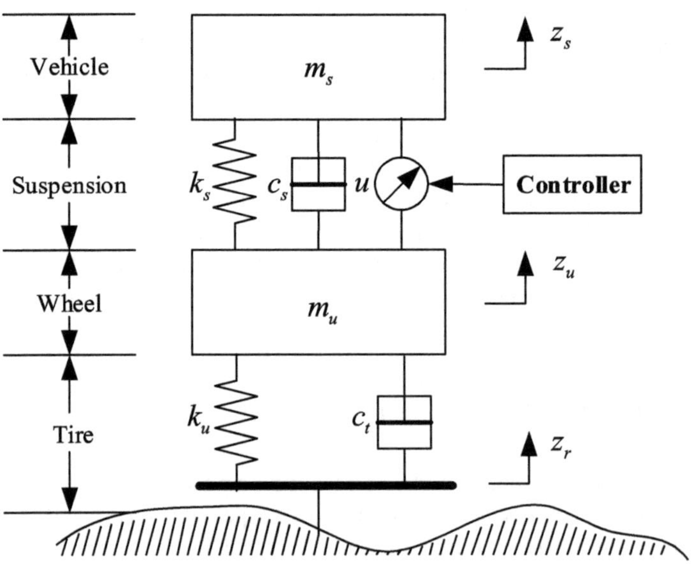

Fig. 1.2 Quarter-Car active suspension model (relative displacement)

compression movements, and the spring stiffness coefficient also holds the nonlinear characteristics. The nonlinear modeling and the corresponding control strategies will be given at the following chapters.

1.2.2 Half-Car Model

If the left and right side of the car are symmetrical, the suspension model can be simplified as a half-car model. Compared to the quarter-car model, half-car suspension systems can reflect both vertical and pitch motions. As shown in Fig. 1.3, M and I stand for the mass of the vehicle body and mass moment of inertia for the pitch motions, respectively, and m_f, m_r are the unsprung masses of front, rear, respectively. F_{df}, F_{dr}, F_{sf} and F_{sr} denote the forces produced by the springs and dampers, respectively, and F_{tf}, F_{bf}, F_{tr}, F_{br} are the elasticity force and damping force of the tires. F_l and F_φ are friction forces of suspension components. z_c is the vertical displacement, φ represents the pitch angle, z_1, z_2 are the unsprung mass displacements and z_{o1}, z_{o2} are the road inputs to the related wheel. a, b show the distances of the suspensions to the center of mass of the vehicle body, and u_1, u_2 are the control inputs of the active suspension systems.

The ideal dynamic equations of the sprung and unsprung masses are given by:

$$\begin{aligned} M\ddot{z}_c &= \Psi_1(t) + u_1 + u_2 + F_l, \\ I\ddot{\varphi} &= \Psi_2(t) + au_1 - bu_2 + F_\varphi, \\ m_f\ddot{z}_1 &= F_{sf} + F_{df} - F_{tf} - F_{bf} - u_1, \\ m_r\ddot{z}_2 &= F_{sr} + F_{dr} - F_{tr} - F_{br} - u_2, \end{aligned} \tag{1.6}$$

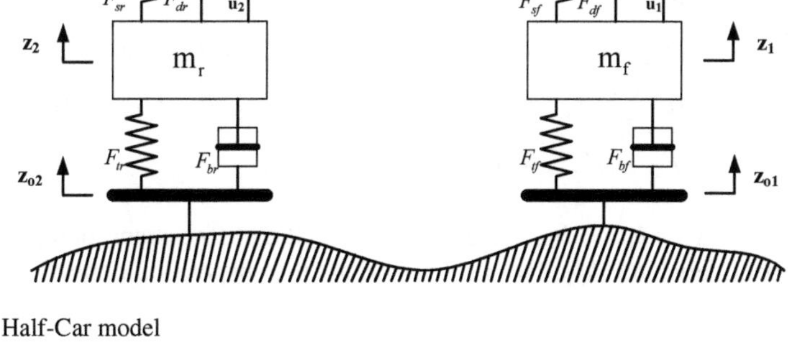

Fig. 1.3 Half-Car model

where

$$\Psi_1(t) = -F_{df} - F_{dr} - F_{sf} - F_{sr},$$
$$\Psi_2(t) = -a(F_{df} + F_{sf}) + b(F_{dr} + F_{sr}).$$

Similar to quarter-car suspensions, the forces produced by springs and dampers can be modeled as linear form and nonlinear form.

1.2.3 Full-Car Model

The full-car suspension model has seven DOF due to the heave, pitch and roll motions of the sprung mass and the vertical motions of the unsprung masses, as depicted in Fig. 1.4. It consists of a single sprung mass (vehicle body) connected to four unsprung masses (front-right, front-left, rear-right and rear-left wheels) at each corner. The sprung mass is free to heave, pitch and roll motions, while the unsprung masses are free to bounce vertically with respect to the sprung mass. The suspension between the sprung mass and the unsprung masses are modeled as dampers and spring elements, while the tires are modeled simple linear springs without damping.

In Fig. 1.4, M, I_x and I_y stand for the mass of the vehicle body, mass moment of inertia for the roll and pitch motions, respectively, and m_i, $i = 1, 2, 3, 4$ are the unsprung masses of front left, front right, rear left, and rear right, respectively. F_{si} and F_{di} denote the forces produced by the spring and damper, respectively, and k_{ti} is the stiffness of the tire. For the vehicle body, z, θ and φ represent the heave, pitch and roll motions, respectively. y_i is the unsprung mass displacement and y_{oi} is the road input to the related wheel. The actuators are placed parallel to the suspension springs and dampers, and their output forces are denoted by u_i. a, b, c and d show the distances of the suspension to the center of mass of the vehicle body. V is the velocity of the vehicle in x-direction.

The dynamic equations of motion for the full vehicle model are obtained as follows, based on an assumption that the pitch and roll angles are small.

$$\Sigma_I \begin{cases} \ddot{z} = -\frac{1}{M}\sum\limits_{i=1}^{4}(F_{di} + F_{si}) + \frac{1}{M}\sum\limits_{i=1}^{4} u_i(t), \\ \ddot{\theta} = -\frac{1}{I_y}(a\sum\limits_{i=1}^{2}(F_{di} + F_{si} - u_i(t)) - b\sum\limits_{i=3}^{4}(F_{di} + F_{si} - u_i(t))), \\ \ddot{\varphi} = -\frac{1}{I_x}(d\sum\limits_{i=2,4}(F_{di} + F_{si} - u_i(t)) - c\sum\limits_{i=1,3}(F_{di} + F_{si} - u_i(t))), \\ \ddot{y}_i = \frac{1}{m_i}\left\{F_{di} + F_{si} - k_{ti}(y_i - y_{oi}) - u_i(t)\right\}, \end{cases} \tag{1.7}$$

where $F_{si} = k_i \Delta y_i$, $F_{di} = b_i \Delta \dot{y}_i$, and k_i, b_i are the stiffness and damping coefficients, respectively, and Δy_i, $i = 1...4$, stand for the suspension deflections.

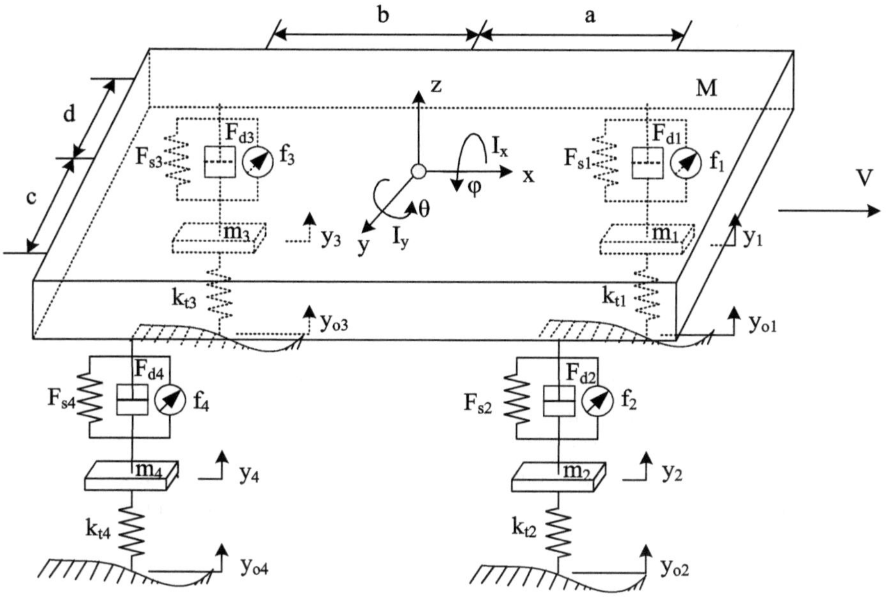

Fig. 1.4 The model of full-car active suspension system

1.3 Road Excitation Model

The load spectrum is typically bandwidth-limited to lower frequencies, and its effects can be relatively easily checked (mostly w.r.t. suspension deflections.) The most relevant for ride studies are ground input disturbances caused by road roughness. There are many possible ways to analytically describe the road inputs, which can be classified as shock or vibration [10]. Shocks are discrete events of relatively short duration and high intensity, as, e.g. caused by a pronounced bump or pothole on an otherwise smooth road. Vibrations, on the other hand, are characterized by prolonged and consistent excitations that are felt on, say, rough roads. Obviously, a well-designed suspension must perform adequately in a wide range of shock and vibration environments. A simple model of the vertical road displacement $z_r(t)$ resulting from a singular disturbance event is given by

$$z_r(t) = \begin{cases} \frac{A}{2}(1-\cos(\frac{2\pi v}{L}t)), & 0 \le t \le \frac{L}{v} \\ 0, & t > \frac{L}{v} \end{cases} \tag{1.8}$$

where A represents the bump height, L is the bump length, and v is the velocity of the passing vehicle [10]. Taking the derivative of $z_r(t)$, we have

$$\dot{z}_r(t) = \begin{cases} \frac{\pi A v}{L}\sin(\frac{2\pi v}{L}t), & 0 \le t \le \frac{L}{v} \\ 0, & t > \frac{L}{v} \end{cases} \tag{1.9}$$

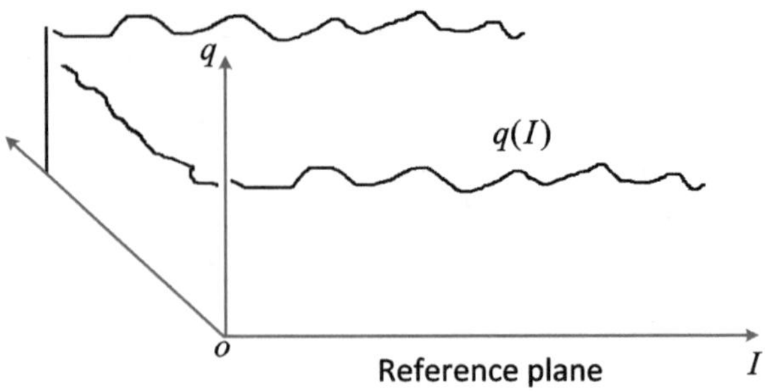

Fig. 1.5 The longitudinal section of the road profile

The longitudinal section of the road profile is shown in Fig 1.5, where q is the height of the road relative to the reference plane along the road lengths I [11]. In the context of vibrations, the road roughness is typically specified as a random process of a given displacement power spectral density (PSD). An often used approximation of measured road displacement PSDs for various terrains is given in the form

$$G(\Omega) = A\Omega^n, \tag{1.10}$$

where Ω is the spatial frequency, typically in unit of radians per length, and A and n are appropriate constants. The most commonly used case corresponds to $n \approx -2$. With this value the displacement spectra of Eq. (1.10) implies a white-noise ground velocity input. The simple expression (1.10) approximates various roads with different degrees of fidelity. The road model fits the white-noise assumption quite well, whereas the comparative fit is less satisfactory for lower frequencies.

The International Standardization Organization (ISO) has proposed a series of standards of road roughness classification using the Power Spectral Density (PSD) values (ISO 2631), as shown in Table 1.1. In addition, the RMS of the road displacement (σ) is used to describe the random road signal power. Due to the ISO, the road displacement PSD can be described as

$$G(n) = G(n_0)\left(\frac{n}{n_0}\right)^{-\omega}, \tag{1.11}$$

Here, n is the space frequency (m^{-1}), n_0 is the reference space frequency, $G(n)$ is the road displacement PSD, $G(n_0)$ is the road displacement PSD under space frequency n_0 which is also called road roughness coefficient shown in Table 1.1, and ω is the linear fitting coefficient which decided the spectrum structure, always $\omega = 2$.

Taking a derivative with respect to the variable $q(I)$, the road velocity PSD can be described as

$$G_{\dot{q}}(n) = (2\pi n)^2 G_q(n) = (2\pi n_0)^2 G_q(n_0) \tag{1.12}$$

From (1.12), the road velocity PSD is a constant over the entire frequency range, which indicates a white-noise input. The amplitude of the road velocity PSD is only

Table 1.1 Road roughness values classified by ISO 2361

Road class	$G(n_0)/(10^{-6}\,\text{m}^3)(n_0 = 0.1\,\text{m}^{-1})$ Geometric mean	$\sigma/(10^{-3}\,\text{m})(0.011\,\text{m}^{-1} < n < 2.83\,\text{m}^{-1})$ Geometric mean
A (very good)	16	3.81
B (good)	64	7.61
C (average)	256	15.23
D (poor)	1024	30.45
E (very poor)	4096	60.90
F	16384	121.80
G	65536	243.61
H	262144	487.22

relevant to road roughness coefficient $G(n_0)$. It can bring much convenience to utilize (1.12) to analyze vibration responses.

Road displacement PSD under time frequency is used to calculate in most cases. Conversion relationship between the space and time frequency is

$$f = nv \tag{1.13}$$

where f is time frequency and v is the vehicle speed. With self correlation function and power spectrum density being a Fourier transform pair, road displacement PSD under spatial frequency can be described as

$$G_q(n) = \int_{-\infty}^{\infty} R(\varsigma)e^{-j2\pi n\varsigma}d\varsigma \tag{1.14}$$

where ς is the distance between two points along the road, similar to time interval τ in self-correlation function $R(\tau)$ in time domain, and we have

$$\varsigma = v\tau \tag{1.15}$$

Substituting the Eqs. (1.13) and (1.15) into the expression (1.14), we have

$$\begin{aligned} G_q(n) = \textstyle\int_{-\infty}^{\infty} R(\varsigma)e^{-j2\pi n\varsigma}d\varsigma = \int_{-\infty}^{\infty} R(v,\tau)e^{-j2\pi\frac{f}{v}v\tau}dv\tau) \\ = v\textstyle\int_{-\infty}^{\infty} R(\tau)e^{-j2\pi f\tau}d(\tau) \end{aligned} \tag{1.16}$$

In (1.16), if the vehicle speed v remains constant, self correlation function $R(v,\tau)$ is only a function of time interval τ, so $R(v,\tau)$ can be replaced by $R(v)$. Rearranging (1.16), (1.17) is obtained.

$$G_q(f) = G_q(n)/v \tag{1.17}$$

By substituting (1.12) and (1.13) to the Eq. (1.17), road displacement PSD under time frequency can be expressed as

$$G_q(f) = G_q(n_0)v(\frac{n_0}{f})^{\omega} \tag{1.18}$$

And the road velocity PSD under time frequency is

$$G_{\dot{q}}(f) = (2\pi f)^2 G_q(f) = 4\pi^2 G_q(n_0) v n_0^2 \tag{1.19}$$

According to the analysis above, the road displacement velocity PSD under time frequency $G_q(f)$ and velocity PSD under time frequency $G_{\dot{q}}(f)$ are both proportional to road roughness coefficient $G(n_0)$ and the vehicle speed v. When v is a fixed value, $G_{\dot{q}}(f)$ can be viewed as a white-noise input.

1.4 Control Objectives of Active Suspension Systems

The essential function of the vehicle suspension is to connect the vehicle body with the wheels. Thereby it is possible to carry the body along the drive way and to transmit forces in the horizontal plane. The suspension gives the wheel a primary vertically aligned movement possibility. As a result, the wheel follows a route with uneven road surfaces to a certain extent. By using spring and damping elements, the resulting body movements are reduced and driving safety and comfort are ensured.

Furthermore, the vehicle suspension influences the position of the wheel relative to the road by its geometry and the spring and damping rate. This allows a systematic influence on the dynamic driving characteristics of the vehicle. The adjustment of these characteristics takes up a compromise, because the requirements of a good driving behavior and a high comfort are the most time inconsistent with one another. Therefore, in designing the control law for a suspension system, usually we need to take the following aspects into consideration [12–14]:

- Ride comfort: it is well-known that ride comfort is an important performance for vehicle design, which is usually evaluated by the body acceleration in the vertical, longitudinal and lateral directions.
- Road holding ability: in order to ensure a firm uninterrupted contact of wheels to road, the dynamic tire load should not exceed the static ones [13].
- Maximum suspension deflection: because of the constraint of mechanical structure, the maximum allowable suspension strokes have to be taken into consideration to prevent excessive suspension bottoming, which can possibly result in deterioration of ride comfort and even structural damage.
- Saturation effect of the actuator: in view of the limited power of the actuator, the control force for the suspension system should be confined to a certain range.

- Reliability of closed-loop systems: the closed-loop systems should be reliable when meeting with non-ideal situations caused by actuators, such as the problems of actuator input delay, sampled data, and fault accommodation for unknown actuator failures.

1.5 Preview of Chapters

This book is made up of seven chapters. In this chapter, we introduce three categories of suspension systems: the two DOF quarter-car model, four DOF half-car model, and seven DOF full-vehicle model. The road excitement model is given in the form of spectrum. In addition, control objectives of active suspension systems are illustrated. The previews of Chaps. 2–8 are as follows.

In Chap. 2, constrained H_∞ control for active suspensions is investigated. The entire frequency H_∞ control scheme is first proposed by solving a convex optimization problem with linear matrix inequality (LMI) constraints. Considering that the vehicle body mass changes with the vehicle load) and its value can be measured online, a load-dependent controller design approach to solve the problem of multi-objective control for vehicle active suspension systems is proposed. The gain matrix depends on the online available information of the body mass based on parameter-dependent Lyapunov functions and the proposed load-dependent approach can yield much less conservative results.

Because human body is much sensitive to vibrations in a certain frequency range, in Chap. 3, we suggest the finite frequency H_∞ controller based on the generalized Kalman-Yakubovich-Popov (KYP) lemma, to achieves better disturbance attenuation performance for the chosen frequency range. Furthermore, according to the online availability of state measurements, dynamic output feedback control problem is solved. All these proposed approaches can provide a good ride comfort, and meantime the constraints required by the real situation are guaranteed in the controller design.

In Chap. 4, constrained active suspension control via nonlinear feedback technology is investigated. Spring nonlinearity and the piece-wise linear behavior of the damper are taken into consideration to form the basis of accurate control. We present the approach to solve the problem of finite-time stabilization for vehicle suspension systems with hard constraints based on terminal sliding-mode (TSM) control. A chattering-free TSM control scheme for suspension systems is proposed, which allows both the chattering and singularity problems to be resolved. Next, an adaptive backstepping control strategy for vehicle active suspensions is presented which is designed to stabilize the attitude of vehicle and meanwhile improve ride comfort in the presence of parameter uncertainties.

In Chap. 5, the problem of actuator saturation is taken into consideration. First, in response to uncertainties in systems and the possible actuator saturation, a saturated adaptive robust control (ARC) strategy is proposed. Specifically, an anti-windup block is added to adjust the control strategy in a manner conductive to stability and performance preservation in presence of saturation. The proposed saturated ARC approach is applied to the half-car active suspension systems, where nonlinear springs and piece-wise linear dampers are adopted. Next, parameter uncertainties, external disturbances, actuator saturation and performance constraints are considered in an unified framework. A constrained adaptive robust control technology is proposed to not only stabilize the attitude of vehicle in the context of parameter uncertainties and external disturbances, but also cover the problems of actuator saturation and performance constraints.

In Chap. 6, we do some researches on reliability of the closed-loop suspension systems. First, the finite frequency method is developed to deal with the problem of suspension control with actuator input delay. After that, robust sampled-data H_∞ control for active vehicle suspension systems is proposed. By using an input delay approach, the active vehicle suspension system with sampling measurements is transformed into a continuous-time system with a delay in the state. The controller design can be solved by linear matrix inequalities (LMIs). Moreover, a fault tolerant control approach is proposed to deal with the problem of fault accommodation for unknown actuator failures of active suspension systems. And we design an adaptive robust controller to adapt and compensate the parameter uncertainties, external disturbances and uncertain nonlinearities generated by the system itself and actuator failures.

In Chaps. 2–6, dynamic process of actuators is neglected when designing the control method. In Chap. 7, actuator dynamics is mainly focused on where electro-hydraulic systems are chosen as actuators. The electro-hydraulic systems are highly non-linear and there exist model uncertainties when establishing the mathematical model, which results in the complexity to design the controller and difficulties of active force tracking. To solve these problems, for the nonlinear half-car model, a filter-based adaptive vibration control scheme is proposed. Furthermore, an adaptive robust vibration control scheme of full-car active suspensions is proposed.

In Chap. 8, in order to reduce the power consumption of active suspensions, an energy-regenerative active suspension control scheme is investigated. Linear DC motors are chosen as the actuators to implement active control. By analyzing energy flow and energy conversion in the DC motors, the criterion of self-powered suspension is presented which can be employed to judge whether an active suspension can be self-powered or not. The energy regeneration implementation scheme is designed to make sure the related physical process of the self-powered suspension is available, where specific operating circuits are illustrated. The energy implementation regeneration scheme is applied into a certain active suspension, and simulation results show that the suspension can be self-powered with acceptant ride comfort.

References

1. D. Hrovat, Survey of advanced suspension developments and related optimal control applications. Automatica **33**(10), 1781–1817 (1997)
2. D. Cao, X. Song, M. Ahmadian, Editors perspectives: road vehicle suspension design, dynamics, and control. Veh. Syst. Dyn. **49**(1–2), 3–28 (2011)

3. R.S. Sharp, S.A. Hassan, The relative performance capabilities of passive, active and semi-active car suspension systems. Proc. Inst. Mech. Eng. Part D J. Automob. Eng. **200**(3), 219–228 (1986)
4. A.F. Naude, J.A. Snyman, Optimisation of road vehicle passive suspension systems. Part 2. Qualification and case study. Appl. Math. Model. **27**(4), 263–274 (2003)
5. H. Du, K.Y. Sze, J. Lam, Semi-active H control of vehicle suspension with magneto-rheological dampers. J. Sound Vib. **283**(283), 981–996 (2005)
6. J.J.H. Paulides, L. Encica, E. Lomonova et al., Design considerations for a semi-active electromagnetic suspension system. IEEE Trans. Magn. **42**(10), 3446–3448 (2006)
7. H. Li, H. Liu, H. Gao et al., Reliable fuzzy control for active suspension systems with actuator delay and fault. IEEE Trans. Fuzzy Syst. **20**(2), 342–357 (2012)
8. G. Priyandoko, M. Mailah, H. Jamaluddin, Vehicle active suspension system using skyhook adaptive neuro active force control. Mech. Syst. Signal Process. **23**(3), 855–868 (2009)
9. C. Poussot-Vassal, C. Spelta, O. Sename et al., Survey and performance evaluation on some automotive semi-active suspension control methods: a comparative study on a single-corner model. Annu. Rev. Control. **36**(1), 148–160 (2012)
10. T.J. Gordon, C. Marsh, M.G. Milsted, A comparison of adaptive LQG and nonlinear controllers for vehicle suspension systems. Veh. Syst. Dyn. **20**(6), 321–340 (1991)
11. Changcheng Zhou, *Vehicle Ride Comfort and Suspension System Design* (China Machine Press, Beijing, 2011)
12. H. Du, N. Zhang, L. Wang, Switched control of vehicle suspension based on motion-mode detection. Veh. Syst. Dyn. **52**(1), 142–165 (2014)
13. Xu Guiqing, Research status and development trend of vehicle suspension. China High-Tech Enterp. **22**, 36 (2010)
14. H. Chen, K.H. Guo, Constrained H_∞ control of active suspensions: an LMI approach. IEEE Trans. Control. Syst. Technol. **13**(3), 412–421 (2005)

Chapter 2
Constrained H_∞ Control for Active Suspensions

In this chapter, the quarter-car model is mainly adopted to develop H_∞ optimal control method. On account of its good disturbance attenuation performances and strong robustness, H_∞ control scheme is relatively efficient to be utilized in active suspensions. In Sect. 2.1, a traditional H_∞ control method in the entire frequency domain is elaborated. The minimum H_∞ norm from the disturbance of the closed-loop system to the vehicle body acceleration is searched for by convex optimal method in order to get the controller gain with the best disturbance attenuation ability and satisfy corresponding performance constraints. In Sect. 2.2, a load-dependent controller design approach is presented to solve the problem of multi-objective control for vehicle active suspension systems. It is assumed that the vehicle body mass whose value changes with the vehicle load resides in an interval and can be measured online. This controller gain matrix depends on the online available information of the body mass, is based on a parameter-dependent Lyapunov function and the proposed controller design approach can yield much less conservative results compared with previous approaches that design robust constant controllers in the quadratic framework. The usefulness and advantage of the proposed controller design methodology are demonstrated via numerical simulations.

Notation: For a matrix P, P^{T}, P^*, P^{-1} and $P^{\perp}$ denote its transpose, conjugate transpose, inverse and orthogonal complement, respectively; the notation $P > 0\,(\geq 0)$ means that P is real symmetric and positive definite (semi-definite); and $[P]_s$ means $P + P^{\mathrm{T}}$. $\|G\|_\infty$ denotes the H_∞-norm of transfer function matrix $G(s)$. For matrices P and Q, $P \otimes Q$ means the Kronecker product. In symmetric block matrices or complex matrix expressions, we use an asterisk $(*)$ to represent a term that is induced by symmetry and diag$\{\ldots\}$ stands for a block-diagonal matrix. Matrices, if their dimensions are not explicitly stated, are assumed to be compatible for algebraic operations. The space of square-integrable vector functions over $[0, \infty)$ is denoted by $L_2[0, \infty)$, and for $w = \{w\,(t)\} \in L_2[0, \infty)$, its norm is given by $\|w\|_2 = \sqrt{\int_{t=0}^{\infty} |w\,(t)|^2\, dt}$.

W. Sun et al., *Advanced Control for Vehicle Active Suspension Systems*, Studies in Systems, Decision and Control 204,
https://doi.org/10.1007/978-3-030-15785-2_2

2.1 Entire Frequency H_∞ Control for Active Suspensions

In this section, considering the output and control constraints, we propose a constrained H_∞ control scheme in the entire frequency domain for active suspensions. The main goal of the constrained H_∞ control is to reduce the vehicle body acceleration as much as possible, which indicates good ride comfort. When designing the controller, time-domain constraints should be taken into consideration. The constraints represent the requirements for good road holding ability, suspension mechanical structure limitation, and avoidance of actuator actuator saturation, which is described in time domain. The state feedback H_∞ controller is solved in the framework of linear matrix inequality (LMI) optimization. Applying the control method to a specific quarter-car model, we can conclude that this approach can realize good ride comfort and limit the time-domain constraints in the given bounds, through the simulation results.

2.1.1 Performance and Time-Domain Constraints of the Active Suspension

In Chap. 1, it is introduced that ride comfort is a remarkably significant performance as far as suspension control is concerned. The main goal of suspension control is to depress the influence of road disturbance as much as possible to make drivers and passengers feel comfortable while driving on the rough road. In the meantime, good road holding ability, maximum suspension deflection, saturation effect of actuator should be taken into account when designing active suspension controller. However, there are conflicts between these control objectives. For example, pursuing too much ride comfort may result in a larger suspension stroke which destroys suspension mechanical structure, excessive suspension bottoming as well as actuator saturation. Therefore, there exists a trade-off among these performances in suspension control.

As a work-around, the active suspension control problem can be described as a disturbance attenuation problem with time-domain constraints [1]. The body acceleration in the vertical direction which represents ride comfort is regarded as the main performance to be optimized. Meanwhile, suspension deflection, dynamic tire load, active control force should be limited in a given bound and they can be viewed as time-domain constraints.

The quarter-car model shown in Fig. 1.2 is considered in this chapter. This model has been used extensively in the literature and captures many important characteristics of more detailed models. Also, the effect of actuator dynamics is neglected and the actuator is modelled as an ideal force generator.

Define the following state variables:

$$x_1(t) = z_s(t) - z_u(t),\ x_2(t) = z_u(t) - z_r(t),\ x_3(t) = \dot{z}_s(t),\ x_4(t) = \dot{z}_u(t),$$

where $x_1(t)$ denotes the suspension deflection, $x_2(t)$ is the tire deflection, $x_3(t)$ is the sprung mass speed, and $x_4(t)$ denotes the unsprung mass speed. We define the disturbance inputs as $w(t) = \dot{z}_r(t)$. Then, by defining $x(t) = \begin{bmatrix} x_1(t) & x_2(t) & x_3(t) & x_4(t) \end{bmatrix}^T$, and according to the dynamic characteristic of the active suspension system (1.5), the state-space form can be given:

$$\dot{x}(t) = Ax(t) + B_w w(t) + Bu(t), \tag{2.1}$$

where

$$A = \begin{bmatrix} 0 & 0 & 1 & -1 \\ 0 & 0 & 0 & 1 \\ -\frac{k_s}{m_s} & 0 & -\frac{c_s}{m_s} & \frac{c_s}{m_s} \\ \frac{k_s}{m_u} & -\frac{k_u}{m_u} & \frac{c_s}{m_u} & -\frac{c_s+c_t}{m_u} \end{bmatrix}, \ B = \begin{bmatrix} 0 \\ 0 \\ \frac{1}{m_s} \\ -\frac{1}{m_u} \end{bmatrix}, \ B_w = \begin{bmatrix} 0 \\ -1 \\ 0 \\ \frac{c_t}{m_u} \end{bmatrix}. \tag{2.2}$$

The velocity of uneven road is defined as the disturbance input:

$$w = \dot{z}_r. \tag{2.3}$$

In order to satisfy the performance requirements, let the car body acceleration z_1 be the measurement output.

Suspension deflection affects vehicle body attitude, and mechanical structure of the suspension also limits the suspension deflection. Excessive suspension bottoming should be avoided, otherwise ride comfort will sharply fall and even destroy the suspension structure. So we require

$$|z_s(t) - z_u(t)| < x_{1\,\max}. \tag{2.4}$$

where $x_{1\,\max}$ is the maximum suspension deflection.

In order to guarantee car safety, we should ensure the firm uninterrupted contact of wheels to road, and the dynamic tire load should be small, that is $k_t(z_u(t) - z_r(t)) < (m_s + m_u)g$. The tire deflection should satisfy the inequality:

$$z_u(t) - z_r(t) < (m_s + m_u)g/k_t = x_{2\,\max}, \tag{2.5}$$

where $x_{2\,\max}$ is the maximum tire deflection.

Another hard constraint imposed on active suspensions is from the limited power of the actuator, that is

$$|u(t)| \le u_{\max}, \tag{2.6}$$

where $u_{\max}$ is the maximum active force.

In order to satisfy the performance requirements, the controlled outputs are defined as

$$z_1(t) = \ddot{z}_s(t), \ z_2(t) = [\, \tfrac{z_s(t)-z_u(t)}{x_{1\,\max}} \ \ \tfrac{z_u(t)-z_r(t)}{x_{2\,\max}} \,]^T. \tag{2.7}$$

Therefore, the vehicle suspension control system can be described by

$$\begin{aligned} \dot{x}(t) &= Ax(t) + Bu(t) + B_w w(t), \\ z_1(t) &= C_1 x(t) + D_1 u(t), \\ z_2(t) &= C_2 x(t), \end{aligned} \tag{2.8}$$

where A, B_w and B are defined in (2.2), and

$$C_1 = \left[-\frac{k_s}{m_s}\ 0\ -\frac{c_s}{m_s}\ \frac{c_s}{m_s} \right],\ C_2 = \begin{bmatrix} \frac{1}{z_{\max}} & 0 & 0\,0 \\ 0 & \frac{k_t}{(m_s+m_u)g} & 0\,0 \end{bmatrix},\ D_1 = \frac{1}{m_s}.$$

2.1.2 Constrained H_∞ Scheme

Our goal is to design a controller to minimize the H_∞ norm from the disturbance input w to the measurement output z_1 and guarantee these time-domain constraints, which means the closed-loop system should satisfy the following equations:

$$\left\| G_{w \to z_1}(j\omega) \right\|_\infty < \gamma, \tag{2.9}$$

where $G_{w \to z_1}(s)$ is called the transfer function from the disturbance input w to the measurement output z_1. Meanwhile, these inequalities should also be respected:

$$|u(t)| \le u_{\max},\ \ |\{z_2(t)\}_i| \le 1,\ i = 1, 2. \tag{2.10}$$

All the state variables are assumed to be observable, and the state feedback controller is presented as

$$u(t) = Kx(t) \tag{2.11}$$

where K is the state feedback gain matrix. Substitute (2.11) into (2.8), we get the state space expression of the closed-loop system:

$$\begin{cases} \dot{x}(t) = (A + BK)x(t) + B_w w(t) \\ z_1(t) = (C_1 + D_1 K)x(t) \\ z_2(t) = C_2 x(t) \end{cases} \tag{2.12}$$

And (2.12) can be abbreviated to

$$\begin{cases} \dot{x}(t) = \bar{A}x(t) + \bar{B}w(t) \\ z_1(t) = \bar{C}x(t) + \bar{D}w(t) \\ z_2(t) = C_2 x(t) \end{cases} \tag{2.13}$$

where

$$\bar{A} = A + BK,\ \bar{B} = B_w,\ \bar{C} = C_1 + D_1 K,\ \bar{D} = 0 \tag{2.14}$$

Based on the analysis above, the process of the controller design is elaborated now. The main procedure is to express (2.9) and (2.10) in the form of linear matrix inequalities (LMIs) [2], and obtain the state feedback gain matrix K through solving the LMIs.

First, we introduce the Schur Complement Lemma:

Lemma 2.1 (Schur Complement Lemma) *Let S be a symmetric matrix given by $S = \begin{bmatrix} S_{11} & S_{12} \\ S_{21} & S_{22} \end{bmatrix}$, and S_{11} is a $r \times r$ matrix, then the following statements are equivalent:*

$$\begin{aligned} &(1) S < 0 \\ &(2) S_{11} < 0,\ S_{22} - S_{12}^T S_{11}^{-1} S_{12} < 0 \\ &(3) S_{22} < 0,\ S_{11} - S_{21}^T S_{22}^{-1} S_{21} < 0 \end{aligned} \tag{2.15}$$

Theorem 2.2 *Let Q be a symmetric positive definite matrix, P be a general matrix, and positive scalars γ, and ρ be given. Matrix C_2 is expressed as $C_2 = \begin{bmatrix} C_{21} \\ C_{22} \end{bmatrix}$. If matrixes P, Q satisfy:*

$$\begin{bmatrix} QA^T + AQ + BY + Y^T B^T & B_w & QC_1^T + Y^T D_1^T \\ * & -\gamma^2 I & 0 \\ * & * & -I \end{bmatrix} < 0 \tag{2.16}$$

$$\begin{bmatrix} -I & \sqrt{\rho} C_{21} Q \\ * & -Q \end{bmatrix} < 0 \tag{2.17}$$

$$\begin{bmatrix} -I & \sqrt{\rho} C_{22} Q \\ * & -Q \end{bmatrix} < 0 \tag{2.18}$$

$$\begin{bmatrix} -I & \sqrt{\rho} Y \\ * & -Q u_{\max}^2 \end{bmatrix} < 0 \tag{2.19}$$

there exists state feedback controller $u(t) = Kx(t)$, such that the closed-loop system in (2.13) satisfies the following conditions:
(1) the closed-loop system is asymptotically stable with $w(t) = 0$;
(2) the H_∞ norm from the disturbance input w to the measurement output z_1 satisfies $\left\| G_{w \to z_1}(j\omega) \right\|_\infty < \gamma$;
(3) the constraints in (2.10) are guaranteed with the disturbance energy under the bound $w_{\max} \le (\rho - V(0))/\gamma^2$.
Moreover, if inequalities (2.16)–(2.19) have a set of feasible solutions, the control gain K is given by

$$K = YQ^{-1} \tag{2.20}$$

Proof. We define $P = Q^{-1}$, $J = \text{diag}\{P, I, I\}$. Performing the congruence transformation to inequality (2.16) by J, inequality (2.16) can be transformed to the following inequality:

$$\begin{bmatrix} A^T P + PA + PBK + K^T B^T P & PB_w & C_1^T + K^T D_1^T \\ * & -\gamma^2 I & 0 \\ * & * & -I \end{bmatrix} < 0. \quad (2.21)$$

Combining (2.14), (2.21) can be written as

$$\begin{bmatrix} \bar{A}^T P + P\bar{A} & P\bar{B} & \bar{C}^T \\ * & -\gamma^2 I & 0 \\ * & * & -I \end{bmatrix} < 0. \quad (2.22)$$

From (2.22), we obviously have

$$\bar{A}^T P + P\bar{A} < 0. \quad (2.23)$$

Let the Lyapunov function of the closed-loop system be

$$V(t) = x^T(t)Px(t). \quad (2.24)$$

Take the derivative of $V(t)$, and if $w(t) = 0$, we have $\dot{x}(t) = Ax(t)$, then $\dot{V}(t)$ can be indicated as

$$\dot{V}(t) = x^T(t)(\bar{A}^T P + P\bar{A})x(t). \quad (2.25)$$

We can determine $\dot{V}(t) < 0$ according to (2.23), which guarantees that the closed-loop system is asymptotically stable with $w(t) = 0$.

By using Schur complement, inequality (2.22) can be written as

$$\begin{bmatrix} \bar{A}^T P + P\bar{A} + \bar{C}^T\bar{C} & P\bar{B} \\ * & -\gamma^2 I \end{bmatrix} < 0, \quad (2.26)$$

and it is equivalent to

$$\begin{bmatrix} x \\ w \end{bmatrix}^T \begin{bmatrix} \bar{A}^T P + P\bar{A} + \bar{C}^T\bar{C} & P\bar{B} \\ * & -\gamma^2 I \end{bmatrix} \begin{bmatrix} x \\ w \end{bmatrix} < 0. \quad (2.27)$$

Expanding (2.27), we get the following inequality

$$x^T(\bar{A}^T P + P\bar{A})x + x^T P\bar{B}w + w^T\bar{B}^T Px + (\bar{C}x)^T\bar{C}x < \gamma^2 w^T w. \quad (2.28)$$

The inequality (2.28) can be transformed to the following form

$$\dot{V} + z_1^T z_1 < \gamma^2 w^T w. \quad (2.29)$$

Integrate both sides of the inequality (2.29) from 0 to ∞, we get

$$\int_0^\infty \dot{V}dt + \int_0^\infty z_1^T z_1 dt < \gamma^2 \int_0^\infty w^T w dt. \tag{2.30}$$

Considering $V(0) = 0,\ V(\infty) = \infty$, the inequality (2.29) is obtained

$$\|z_1\|_2^2 < \gamma^2 \|w\|_2^2, \tag{2.31}$$

which means $\left\|G_{w\rightarrow z_1}(j\omega)\right\|_\infty < \gamma$. Hence, the H_∞ performance of the closed-loop system is satisfied.

It comes to the time-domain constraints. Based on (2.29), we have

$$\dot{V} < \gamma^2 w^T w. \tag{2.32}$$

Integrating both sides of the inequality (2.32) from 0 to t and noting that $w_{\max} \leq (\rho - V(0))/\gamma^2$, (2.33) is shown:

$$V(t) < V(0) + \gamma^2 \int_0^t w^T w dt \leq V(0) + \gamma^2 \|w\|_2^2 \leq \rho. \tag{2.33}$$

Define $J_2 = \mathrm{diag}\{I, P\}$, and we perform a congruence transformation to (2.17), (2.18), (2.19), respectively, by the full rank matrix J_2^T on the left, and J_2 on the right. The results are shown as

$$\begin{bmatrix} -I & \sqrt{\rho}C_{21} \\ * & -P \end{bmatrix} < 0, \tag{2.34}$$

$$\begin{bmatrix} -I & \sqrt{\rho}C_{22} \\ * & -P \end{bmatrix} < 0, \tag{2.35}$$

$$\begin{bmatrix} -I & \sqrt{\rho}K \\ * & -u_{\max}^2 P \end{bmatrix} < 0. \tag{2.36}$$

In terms of the suspension deflection, $|x_1| < x_{1\max}$ is required, that is

$$\max_{t\geq 0} |C_{21}x|^2 < 1. \tag{2.37}$$

Expand the expression $\max\limits_{t\geq 0} |C_{21}x|^2$

$$\begin{aligned} &\max_{t\geq 0} |C_{21}x|^2 = \max_{t\geq 0} \left|x^T C_{21}^T C_{21} x\right| = \max_{t\geq 0} \left|x^T P^{\frac{1}{2}} P^{-\frac{1}{2}} C_{21}^T C_{21} P^{-\frac{1}{2}} P^{\frac{1}{2}} X\right| \\ &\leq \lambda_{\max}(P^{-\frac{1}{2}} C_{21}^T C_{21} P^{-\frac{1}{2}}) x^T P x = \lambda_{\max}(P^{-\frac{1}{2}} C_{21}^T C_{21} P^{-\frac{1}{2}}) V(t) \\ &\leq \lambda_{\max}(P^{-\frac{1}{2}} C_{21}^T C_{21} P^{-\frac{1}{2}})\rho. \end{aligned} \tag{2.38}$$

By using Schur complement, inequality (2.34) is equivalant to

$$\rho P^{-\frac{1}{2}} C_{21}^T C_{21} P^{-\frac{1}{2}} < I. \tag{2.39}$$

Multiplying the inequality (2.34) by $P^{-\frac{1}{2}}$ from both left and ride sides leads to

$$\rho P^{-\frac{1}{2}} C_{21}^T C_{21} P^{-\frac{1}{2}} < x_{1\max}^2 I, \tag{2.40}$$

which meets the requirement of (2.38). Hence, if the disturbance energy is under the bound $w_{\max} \leq (\rho - V(0))/\gamma^2$, the suspension deflection can be limited in the given range. In the same way, with regard to the tire deflection and the control input u, we have

$$\begin{aligned} \max_{t\geq 0} |C_{22}x|^2 &\leq \lambda_{\max}(P^{-\frac{1}{2}} C_{21}^T C_{21} P^{-\frac{1}{2}})\rho, \\ \max_{t\geq 0} |Kx|^2 &\leq \lambda_{\max}(P^{-\frac{1}{2}} K^T K P^{-\frac{1}{2}})\rho. \end{aligned} \tag{2.41}$$

Inequalities (2.35) and (2.36) can be transformed into the following forms

$$\begin{aligned} \rho P^{-\frac{1}{2}} C_{22}^T C_{22} P^{-\frac{1}{2}} &< I \\ \rho P^{-\frac{1}{2}} K^T K P^{-\frac{1}{2}} &< u_{\max}^2 I \end{aligned} \tag{2.42}$$

So we have

$$\begin{aligned} \max_{t\geq 0} |C_{22}x|^2 &< 1 \\ \max_{t\geq 0} |Kx|^2 &< u_{\max}^2 \end{aligned} \tag{2.43}$$

All the three time-domain constraints are guaranteed. The proof is completed.

2.1.3 Simulation Verification

The parameters of the suspension system are listed in Table 2.1, and the maximums of the time-domain constraints are given in Table 2.2.

Table 2.1 Parameters of the suspension system

Sprung mass m_s	320 kg
Unsprung mass m_u	40 kg
Suspension stiffness k_s	18000 N/m
Tire stiffness k_u	200000 N/m
Suspension damping coefficient c_s	1000 Ns/m
Tire damping coefficient c_t	10 Ns/m

Table 2.2 Maximums of the time-domain constraints

The maximum suspension deflection $x_{1\max}$	0.1 m
The maximum tire deflection $x_{2\max}$	0.01764 m
The maximum active force $u_{\max}$	2500 N
The positive scalar ρ	1

According to Theorem 2.2, the H_∞ state feedback gain of the suspension system is figured out by MATLAB, that is

$$K = 10^4 \times [\, 1.0098 \;\; 4.9655 \;\; -0.1896 \;\; 0.0909 \,], \tag{2.44}$$

with the optimal value of γ reaching $\gamma^* = 8.7955$.

Bump Road Inputs

Consider the case of an isolated bump in an otherwise smooth road surface, the road displacement input is described in the form of (1.9). Let $A = 0.08\,\text{m}$, $L = 5\,\text{m}$, $V = 12.5\,\text{m/s}$, the road displacement input can be rewritten as

$$z_r(t) = \begin{cases} 0.04(1 - \cos 5\pi t) & 0 \le t \le 0.4s \\ 0 & t > 0.4s \end{cases} \tag{2.45}$$

The disturbance input signal of the system (1.8) is the sinusoidal signal, that is

$$\dot{z}_r(t) = \begin{cases} 0.2\pi \sin(5\pi t) & 0 \le t \le 0.4s \\ 0 & t > 0.4s \end{cases} \tag{2.46}$$

The simulation results of the active suspension with the designed H_∞ controller are compared with those of passive suspension. The bode diagrams of passive and active suspension are shown in Figs. 2.1 and 2.2 respectively. From Fig. 2.3, the vehicle heave acceleration of the active suspension is obviously smaller than that of the passive suspension. It can be indicated that the active suspension can provide better ride comfort, with lower peak and shorter settling time in the vehicle heave acceleration.

As in Fig. 2.4, the ratios of suspension stroke and the maximum limitation of passive and active suspension are less than 1, meaning that suspension deflection of the passive and active suspension is limited in the given bound. The first peak of the active suspension is a little higher than that of the passive suspension. However, response time of the controlled system is shortened, and fluctuation times are also reduced, which means the suspension deflection can stabilize faster. In terms of tire deflection, the first peak of both the active and passive suspension is nearly the same within the limit, while response time and fluctuation times of controlled system are reduced, which is shown in Fig. 2.5. The active force is described in Fig. 2.6, and we notice that the maximum of the actual force value is 918 N, which is much smaller than 2500 N.

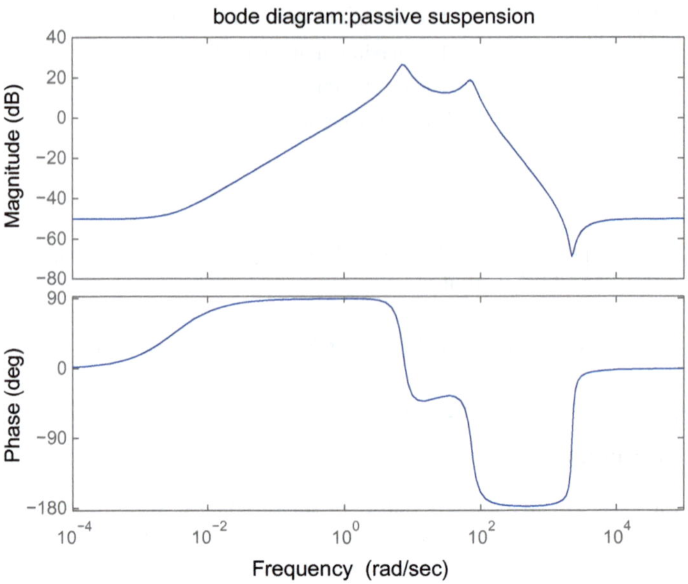

Fig. 2.1 Bode diagram of passive suspension

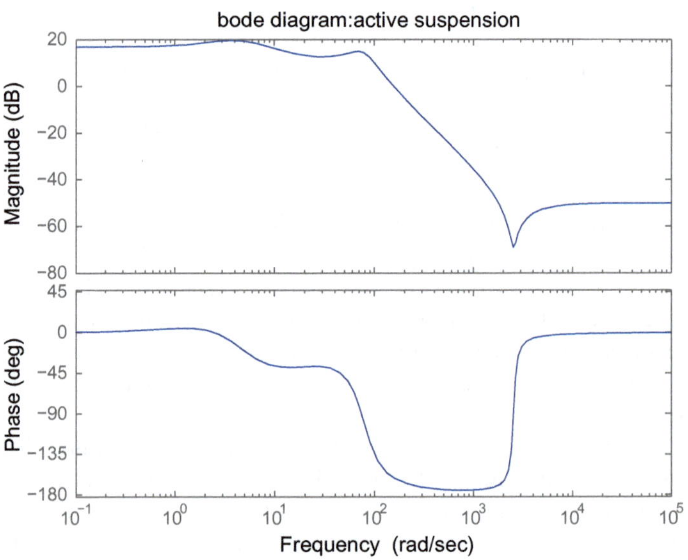

Fig. 2.2 Bode diagram of active suspension

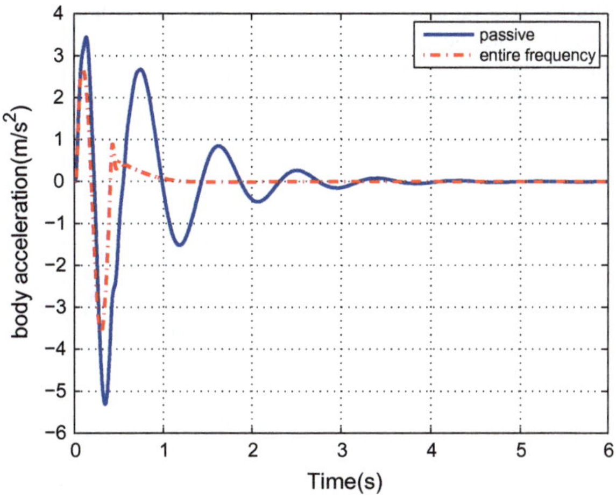

Fig. 2.3 The vehicle heave acceleration of active suspension compared with passive suspension

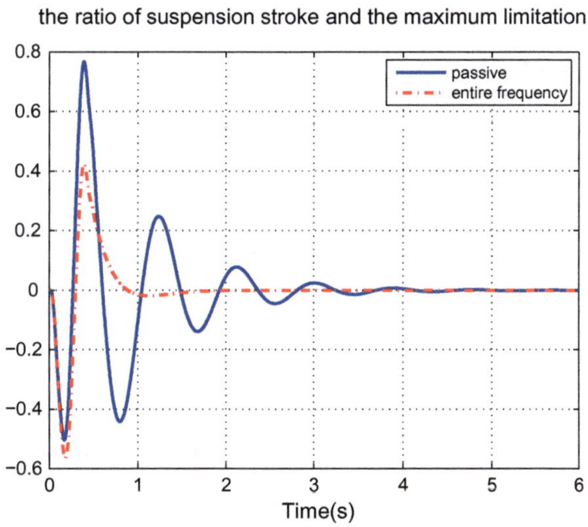

Fig. 2.4 The ratio of suspension stroke and the maximum limitation of active suspension compared with passive suspension

Random Road Inputs

Simulations are conducted under random road excitation in the form of white noise. The power spectra density of uneven road velocity input is described as 1.19 in Chap. 1:

$$G_{\dot{q}}(f) = 4\pi^2 G_q(n_0) v n_0^2.$$

Let the road class be C, with road roughness coefficient of $256 \times 10^{-6}\text{m}^3$, and the vehicle drives at the speed of 45 km/h. The power spectra density of the vehicle heave acceleration can be described as:

Fig. 2.5 The ratio of tire deflection and the maximum limitation of active suspension compared with passive suspension

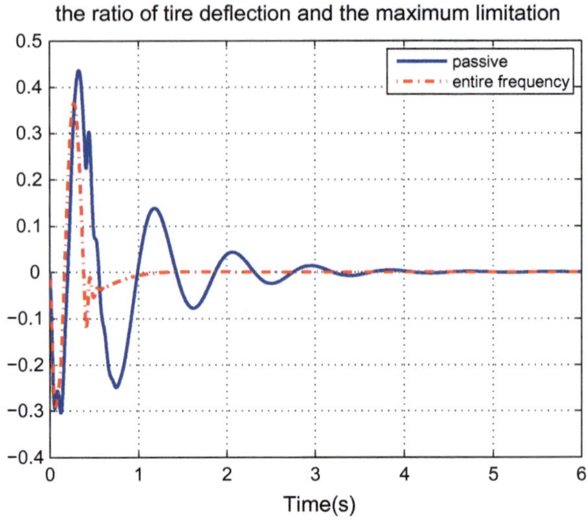

Fig. 2.6 The active control force of active suspension

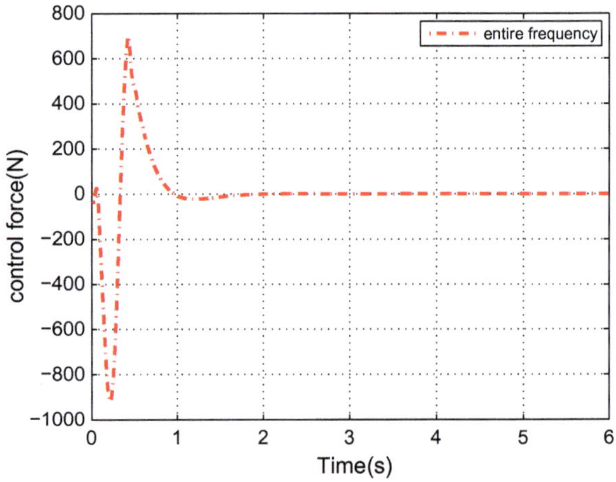

$$\phi_{z_1}(f) = \left|G_{w \to z_1}(j\omega)\right|^2 G_{\dot{q}}(f). \tag{2.47}$$

Based on (2.47), the figure showing PSD body acceleration of active suspension and passive suspension is obtained. From Fig. 2.7, we can see the curve of the active suspension is below that of the passive suspension, especially in low frequency. It can be concluded that ride comfort is remarkably improved when constrained H_∞ control scheme is applied.

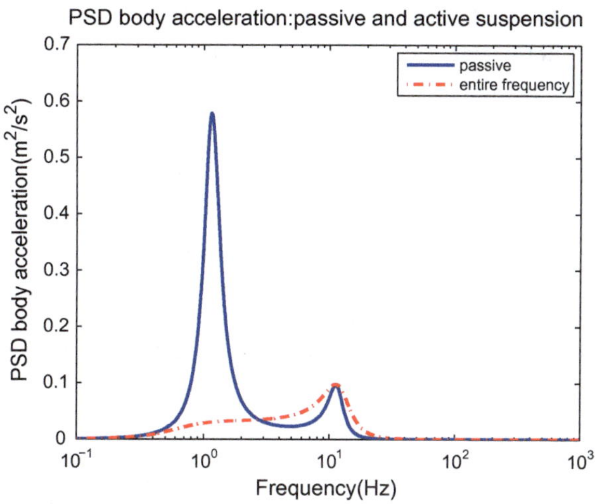

Fig. 2.7 PSD body acceleration of active suspension and passive suspension

2.1.4 Conclusion

In this section, active suspension control problem is formulated as a disturbance attenuation problem with hard time-domain constraints. In the framework of LMI optimization and multiobjective control, a constrained H_∞ control method is proposed. Conflicting requirements are guaranteed in the limits, and ride comfort is obviously improved through convex optimization. Compared with standard methods with a single weighted objective, the proposed control scheme to design active suspensions is simple and elegant. The design procedure is based on a 2 DOF quarter-car model without actuator dynamics. Furthermore, analysis and simulation results have proved that the proposed constrained H_∞ active suspension can maintain better ride comfort, while keeping suspension strokes and control inputs within bounds and ensuring firm contact of wheels to road.

2.2 Load-Dependent Control for Active Suspensions

This section presents a load-dependent controller design approach to solve the problem of multi-objective control for vehicle active suspension systems by using linear matrix inequalities. A quarter-car model with active suspension system is considered in this section. It is assumed that the vehicle body mass resides in an interval and can be measured online. This approach of designing controllers, whose gain matrix depends on the online available information of the body mass, is based on a parameter-dependent Lyapunov function. Since the parameter-dependent idea is fully exploited, the proposed controller design approach can yield much less conservative results compared with previous approaches that design robust constant controllers in the quadratic framework. The usefulness and the advantages of the proposed controller design methodology are demonstrated via numerical simulations.

2.2.1 Problem Formulation

Consider the quarter-car model shown in Fig. 1.1, where m_s is the sprung mass whose value changes with the vehicle load and the damping coefficients of the sprung shock-absorber and the tire of c_t is neglected because its value is usually very small. The ideal dynamic equations for the sprung and unsprung masses of the quarter-car model is given by (1.5), where $c_t = 0$, that is

$$\begin{aligned} &m_s\ddot{z}_s(t) + c_s[\dot{z}_s(t) - \dot{z}_u(y)] + k_s[z_s(t) - z_u(t)] = u(t),\\ &m_s\ddot{z}_u(t)+c_s[\dot{z}_u(t)-\dot{z}_s(y)]+k_s[z_u(t)-z_s(t)]+k_u[z_u(t)-z_r(t)]=-u(t). \end{aligned} \tag{2.48}$$

The disturbance input of the system is the road displacement z_r, which can be represented by (2.48) according to [3, 4]

$$\dot{z}_r(t) = 2\pi q_0\sqrt{G_0 V}w(t), \tag{2.49}$$

where G_0 stands for the road roughness coefficient, q_0 is the reference spatial frequency, V is the vehicle forward velocity, $w(t)$ is zero-mean white noise with identity power spectral density.

As is mentioned previously, the body mass m_s usually changes with the vehicle load. Throughout the section, it is assumed that the vehicle body mass resides in an interval and can be measured online, that is,

$$m_1 \le m_s \le m_2. \tag{2.50}$$

Choose the following set of state variables:

$$\begin{aligned} &x_1(t) = z_s(t) - z_u(t),\ x_2(t) = z_u(t) - z_r(t),\\ &x_3(t) = \dot{z}_s(t),\ x_4(t) = \dot{z}_u(t), \end{aligned} \tag{2.51}$$

where x_1 is the suspension deflection, x_2 is the tire deflection, x_3 is the sprung mass speed, and x_4 is the unsprung mass speed.

Then, by defining $x(t) \triangleq \left[x_1(t)\ x_2(t)\ x_3(t)\ x_4(t)\right]^T$, the dynamic equations in (2.48) can be written in the following state-space form:

$$\dot{x}(t) = A(m_s)x(t) + B(m_s)u(t) + B_w(m_s)w(t), \tag{2.52}$$

where

$$A(m_s) = \begin{bmatrix} 0 & 0 & 1 & -1 \\ 0 & 0 & 0 & 1 \\ -\frac{k_s}{m_s} & 0 & -\frac{c_s}{m_s} & \frac{c_s}{m_s} \\ \frac{k_s}{m_u} & -\frac{k_u}{m_s} & \frac{c_s}{m_s} & -\frac{c_s}{m_s} \end{bmatrix},$$
$$B(m_s) = \begin{bmatrix} 0 \\ 0 \\ \frac{1}{m_s} \\ -\frac{1}{m_u} \end{bmatrix}, \quad B_w(m_s) = \begin{bmatrix} 0 \\ -2\pi q_0\sqrt{G_0 V} \\ 0 \\ 0 \end{bmatrix}. \tag{2.53}$$

It is worth mentioning that as the body mass m_s usually changes with the vehicle load, which can be measured online, we express the system matrices of the quarter-car model as functions of m_s.

In designing the control law for a suspension system, usually we need to take the following aspects into consideration according to Sect. 1.4:

(1) Ride comfort: Ride comfort is an important performance for vehicle design, which is usually evaluated by the body acceleration in the vertical direction. Therefore, in the controller design, one of our main objectives is to minimize the vertical body acceleration $\ddot{z}_s$, that is,

$$\min \ddot{z}_s. \tag{2.54}$$

(2) Road holding ability: In order to ensure a firm uninterrupted contact of wheels to road, the dynamic tire load should not exceed the static ones, that is,

$$k_u(z_u - z_r) < 9.8(m_s + m_u). \tag{2.55}$$

By considering (2.50), (2.55) holds if

$$k_u(z_u - z_r) < 9.8(m_1 + m_u). \tag{2.56}$$

(3) Maximum suspension deflection: The maximum allowable suspension stroke has to be taken into consideration to prevent excessive suspension bottoming, which can possibly result in deterioration of ride comfort and even structural damage. The requirement is

$$|z_s(t) - z_u(t)| \le z_{\max}, \tag{2.57}$$

where $z_{\max}$ is the maximum suspension deflection.

(4) Saturation effect of actuator: In view of the limited power of the hydraulic actuator, the active force for the suspension system should be confined to a certain range, that is,

$$|u(t)| \le u_{\max}. \tag{2.58}$$

The latter three requirements are actually constraints, while only the first one needs to be minimized. The strategy in designing control law for suspension systems is to minimize the vertical body acceleration $\ddot{z}_s$ while keeping the other three requirements satisfied.

According to the above four requirements, we define the following output variables:

$$\begin{aligned} z_1(t) &= \ddot{z}_s, \\ z_2(t) &= (z_s(t) - z_u(t))/z_{\max} \\ z_3(t) &= k_u(z_u - z_r)/9.8(m_s + m_u). \\ z_4(t) &= u(t)/u_{\max}. \end{aligned} \tag{2.59}$$

Therefore, the vehicle suspension system can be described by the following state-space equation:

$$\begin{aligned} \dot{x}(t) &= A(m_s)x(t) + B(m_s)u(t) + B_w(m_s)w(t), \\ z_l(t) &= C_l(m_s)x(t) + D_l(m_s)u(t),\ l = 1, ..., 4, \end{aligned} \tag{2.60}$$

where $A(m_s)$, $B(m_s)$, $B_w(m_s)$ are defined in (2.53), and

$$\begin{aligned} C_1(m_s) &= [-\tfrac{k_s}{m_s}\ 0\ -\tfrac{c_s}{m_s}\ \tfrac{c_s}{m_s}],\ D_1(m_s) = \tfrac{1}{m_s}, \\ C_2(m_s) &= [1/z_{\max}\ 0\ 0\ 0],\ D_2(m_s) = \tfrac{1}{m_s}, \\ C_3(m_s) &= [0\ k_u/9.8(m_1 + m_u)\ 0\ 0],\ D_3(m_s) = 0, \\ C_4(m_s) &= [0\ 0\ 0\ 0],\ D_4(m_s) = \tfrac{1}{u_{\max}}. \end{aligned} \tag{2.61}$$

It is not difficult to see that the system matrices which are dependent on the body mass m_s can be expressed as

$$(A(m_s), B(m_s), B_w(m_s), C_l(m_s), D_l(m_s)) = \sum_{i=1}^{2} \alpha_i(A_i, B_i, B_{wi}, C_{li}, D_{li}),$$

$$\alpha_i \geq 0,\ \alpha_1 + \alpha_2 = 1, \tag{2.62}$$

where

$$\begin{aligned} (A_1, B_1, B_{w1}, C_{l1}, D_{l1}) &= (A(m_s), B(m_s), B_w(m_s), C_l(m_s), D_l(m_s))|_{m_s=m_1}, \\ (A_2, B_2, B_{w2}, C_{l2}, D_{l2}) &= (A(m_s), B(m_s), B_w(m_s), C_l(m_s), D_l(m_s))|_{m_s=m_2}. \end{aligned}$$

Moreover, the relationship between the vector a9ea1; a2T and the online measurable body mass m_s is given by

$$\alpha_1 = (\frac{1}{m_s} - \frac{1}{m_2})/(\frac{1}{m_1} - \frac{1}{m_2}),\ \alpha_2 = (\frac{1}{m_1} - \frac{1}{m_s})/(\frac{1}{m_1} - \frac{1}{m_2}). \tag{2.63}$$

Now we have used a two-vertex polytope to describe the load-dependent system matrices. The polytope description has been used in many references (see, for instance, [5, 6]). As the vehicle load can be measured online easily, we can get the vector (α_1, α_2) according to the available m_s based on (2.63).

For the active suspension system (2.60), [4] designs a state-feedback control law of the following form:

$$u(t) = Kx(t), \tag{2.64}$$

where K is a constant feedback gain to be determined. This robust control approach to achieve multi-objective performances deserves some remarks:

(1) The controller design presented in [4] is based on the notion of quadratic stability. That is, for the entire uncertainty polytope, a fixed Lyapunov function is required to satisfy a set of LMI conditions. Such treatment has been well recognized to be conservative, and an advanced research topic in robust control is to utilize parameter-dependent Lyapunov functions.

(2) In the design example in [4], the considered uncertain parameter is the sprung mass ms, whose value depends on the total value of the load. It is noted that the designed controller has a constant gain matrix K, which is used for all possible values of body mass m_s. In fact, the body mass m_s is usually not difficult to obtain online, thus if the controller gain can change according to the online available value of m_s, better performance may be achieved.

Based on the above points, we consider the following controller structure:

$$u(t) = K(m_s)x(t), \tag{2.65}$$

where $K(m_s)$ is a gain matrix function to be determined. Our purpose is to investigate the design of the controller (2.65) based on parameter-dependent Lyapunov functions. Since the controller design presented here does not belong to those commonly used robust control approaches, and is essentially different from those in the quadratic framework, we call (2.65) a load-dependent controller.

By applying controller (2.65) to the suspension system (2.60), we obtain the following closed-loop system:

$$\begin{aligned} \dot{x}(t) &= \bar{A}(m_s)x(t) + B_w(m_s)w(t), \\ z_l(t) &= \bar{C}_l(m_s)x(t), \ l = 1, ..., 4, \end{aligned} \tag{2.66}$$

where

$$\begin{aligned} \bar{A}(m_s) &\stackrel{\Delta}{=} A(m_s) + B(m_s)K(m_s), \\ \bar{C}_l(m_s) &\stackrel{\Delta}{=} [\bar{C}_l(m_s) + D_l(m_s)K(m_s)], \ l = 1, ..., 4. \end{aligned} \tag{2.67}$$

Then, the transfer functions from the disturbance signal to the controlled outputs are given by

$$T_l(s) = \bar{C}_l(m_s)[sI - \bar{A}(m_s)]^{-1}B_w(m_s), l = 1, ..., 4. \tag{2.68}$$

Similar to [4], we also introduce the H_2 and GH_2 (generalized H_2, also called $L_2 - L_\infty$ [7]) performances to evaluate the controlled outputs $z_j(t)$. Based on the aforementioned requirements for control design, the problem to be solved in this section can be summarized as follows:

Given an active suspension system (2.60), design a load-dependent controller in the form of (2.65) via the following minimization problem:

$$\min \gamma_1 \text{ s.t.} \begin{cases} \|T_1(s)\|_2 \leq \gamma_1, \\ \|T_l(s)\|_G \leq \gamma_l, \quad l = 2, 3, 4, \\ OC, \end{cases}$$

where γ_l is a given constant,

$$\|T_1(s)\|_2 \triangleq \sqrt{\frac{1}{2\pi}\int_{-\infty}^{\infty} \text{Tr } T_1^*(jw)T_1(jw)dw},$$
$$\|T_l(s)\|_G \triangleq \sup\left\{\|z_l(t)\| \,:\, x(0) = 0,\ t \geq 0,\ \int_0^t \|\omega(\tau)\|\mathrm{d}\tau \leq 1\right\}$$

and OC represents other constraints, such as pole constraints of the closed-loop system. The proposed control system diagram is given in Fig. 2.8.

Remark 2.3 It is worth mentioning that the control strategy proposed above is much different from the standard PID control, as seen in the following two aspects:

(1) In the standard PID controller design, it is often difficult to take the variation of the body mass m_s into consideration, which is dependent on the vehicle load. However, in the control strategy proposed above, the body mass m_s is assumed to reside in an interval, which characterizes the real situation more precisely.

(2) In the control strategy proposed above, the multiple requirements (including ride comfort, road holding ability, suspension deflection limit and saturation effect of actuator) are formulated in a unified framework, based on which the controller design is cast into a multiple-objective minimization problem. However, in the standard PID controller design, we are usually difficult to take these factors into account simultaneously.

2.2.2 *Load-Dependent Controller Design*

The problem of multi-objective control through loaddependent controllers formulated in the above section will be investigated. First, according to [2], the closed-loop

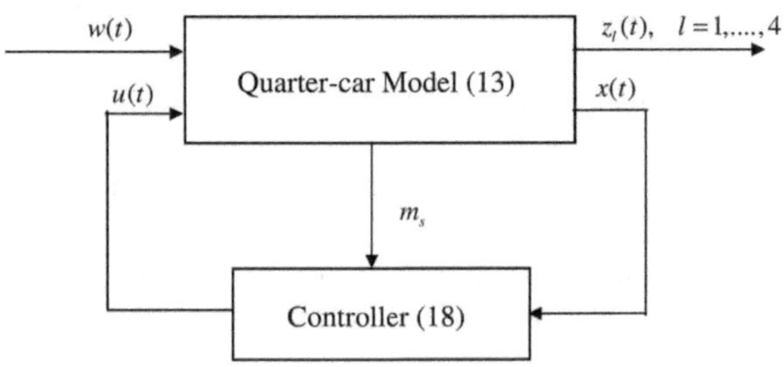

Fig. 2.8 Control system diagram

system in (2.66) is asymptotically stable with $\|T_1(s)\|_2 \leq \gamma_1$ and $\|T_l(s)\|_G \leq \gamma_l$, $l = 2, 3, 4$, if and only if there exist matrix functions $P(m_s)$ and $S(m_s)$ satisfying

$$\mathrm{Tr}(S(m_s)) < \gamma_1^2, \tag{2.69}$$

$$\begin{bmatrix} \bar{A}^T(m_s)P(m_s) + P(m_s)\bar{A}(m_s) & P(m_s)B_w(m_s) \\ * & -I \end{bmatrix} < 0, \tag{2.70}$$

$$\begin{bmatrix} -S(m_s) & \bar{C}_1(m_s) \\ * & -P(m_s) \end{bmatrix} < 0, \tag{2.71}$$

$$\begin{bmatrix} -\gamma_l^2 I & \bar{C}_l(m_s) \\ * & -P(m_s) \end{bmatrix} < 0, \ l = 2, 3, 4. \tag{2.72}$$

In addition, in order to obtain desired dynamics of the closed-loop systems, usually some pole placement constraints need to be imposed. Here, we consider the following two kinds of regional pole constraints [8]:

(1) Disk region: Let $\mho(\eta, \rho)$ denotes any disk region centered in η with radius ρ in the complex plane $\eta, \rho \in \mathcal{R}$ and $\rho > 0$. Then, all the eigenvalues of $\bar{A}(m_s)$ in (2.66) lie in the region $\mho(\eta, \rho)$ if and only if there exists a matrix function $P(m_s) > 0$ satisfying

$$\begin{bmatrix} -P(m_s) & P(m_s)(\bar{A}(m_s) - \eta I) \\ * & -\rho^2 P(m_s) \end{bmatrix} < 0. \tag{2.73}$$

(2) Vertical strip: Let $\wp(\upsilon, \mu)$ denotes a vertical strip lying within the bounds υ and μ ($\upsilon < \mu$, υ, $\mu \in \mathcal{R}$). Then, all the eigenvalues of $\bar{A}(m_s)$ in (2.66) lie in the region $\wp(\upsilon, \mu)$ if and only if there exists a matrix function $P(m_s) > 0$ satisfying

$$(\bar{A}(m_s) - \mu I)^T P(m_s) + P(m_s)(\bar{A}(m_s) - \mu I) < 0, \tag{2.74}$$

$$(\bar{A}(m_s) - \mu I)^T P(m_s) + P(m_s)(\bar{A}(m_s) - \mu I) < 0. \tag{2.75}$$

In the multi-objective synthesis, in order to cast the controller design into convex optimization problems, we usually need to set a common Lyapunov matrix for different performance objectives. Thus, the closed-loop system (2.66) is asymptotically stable with $\|T_1(s)\|_2 \leqslant \gamma_1$ and $\|T_l(s)\|_G \leqslant \gamma_l$, $l = 2, 3, 4$, and all the eigenvalues of $\bar{A}(m_s)$ lie in the region $\mho(\eta, \rho)$ (or $\wp(\upsilon, \mu)$) if there exist matrix functions $P(m_s) > 0$ and $S(m_s) > 0$ satisfying (2.69)–(2.73) (or (2.69)–(2.72), (2.74) and (2.75)). Chen et al. [4] presents a robust controller design by setting $P(m_s) \equiv P$ for the entire uncertainty domain. In the following, we will present a new approach based on parameter-dependent Lyapunov functions.

First define the following invertible matrix functions:

$$\begin{aligned} J_1 &\triangleq \mathrm{diag}\{P^{-1}(m_s), I\}, \\ J_2 &\triangleq \mathrm{diag}\{I, P^{-1}(m_s)\}, \\ J_3 &\triangleq \mathrm{diag}\{P^{-1}(m_s), P^{-1}(m_s)\}. \end{aligned} \tag{2.76}$$

By performing congruence transformations to (2.70)–(2.75) by J_1, J_2, J_2, J_3, $P^{-1}(m_s)$, $P^{-1}(m_s)$ respectively, and by changing the matrix variables with

$$\bar{P}(m_s) \triangleq P^{-1}(m_s),\ \bar{K}(m_s) \triangleq K(m_s)P^{-1}(m_s) \tag{2.77}$$

we obtain

$$\begin{bmatrix} \left[A(m_s)\bar{P}(m_s) + B(m_s)\bar{K}(m_s)\right]_s & B_w(m_s) \\ * & -I \end{bmatrix} < 0, \tag{2.78}$$

$$\begin{bmatrix} -S(m_s) & C_1(m_s)\bar{P}(m_s) + D_1(m_s)\bar{K}(m_s) \\ * & -\bar{P}(m_s) \end{bmatrix} < 0, \tag{2.79}$$

$$\begin{bmatrix} -\gamma I & C_l(m_s)\bar{P}(m_s) + D_l(m_s)\bar{K}(m_s) \\ * & -\bar{P}(m_s) \end{bmatrix} < 0,\ l = 2, 3, 4, \tag{2.80}$$

$$\begin{bmatrix} -\bar{P}(m_s) & (A(m_s) - \eta I)\bar{P}(m_s) + B(m_s)\bar{K}(m_s) \\ * & -\rho^2\bar{P}(m_s) \end{bmatrix} < 0, \tag{2.81}$$

$$\left[(A(m_s) - \mu I)\bar{P}(m_s) + B(m_s)\bar{K}(m_s)\right]_s < 0, \tag{2.82}$$

$$\left[-(A(m_s) - \upsilon I)\bar{P}(m_s) - B(m_s)\bar{K}(m_s)\right]_s < 0. \tag{2.83}$$

Equations (2.69), (2.78)–(2.81) are the conditions for the existence of admissible controllers with disk pole constraint, and (2.69), (2.69), (2.82), (2.83) are the conditions for the existence of admissible controllers with vertical strip pole constraint. From (2.77) we know that if there exist matrix functions $\bar{P}(m_s)$, $S(m_s)$ and $\bar{K}(m_s)$ satisfying the above required conditions, the gain matrix function for an admissible controller in the form of (2.65) can be given by

$$\bar{K}(m_s) \triangleq K(m_s)P(m_s). \tag{2.84}$$

It is noted that for fixed m_s, conditions (2.69), (2.78)–(2.83) are LMIs, which can be readily solved via standard numerical software. However, these conditions cannot be implemented due to their infinite-dimensional nature in the parameter m_s. Our purpose hereafter is to transform these conditions into tractable LMI-based conditions.

According to the inner property of the polytopic uncertain systems, we assume the matrix functions $\bar{P}(m_s)$, $S(m_s)$ and $\bar{K}(m_s)$ in (2.69), (2.78)–(2.83) to be of the following form:

$$\bar{P}(m_s) = \sum_{i=1}^{2} \alpha_i \bar{P}_i, \; S(m_s) = \sum_{i=1}^{2} \alpha_i S_i, \; \bar{K}(m_s) = \sum_{i=1}^{2} \alpha_i \bar{K}_i. \tag{2.85}$$

Then, (2.69) holds if

$$\mathrm{Tr}(S_i) < \gamma_1^2, \; i = 1, 2. \tag{2.86}$$

In addition, it is not difficult to rewrite (2.78) in the following form:

$$\begin{aligned} \mathrm{X}(m_s) \overset{\Delta}{=} & \begin{bmatrix} \mathrm{sym}([A(m_s)\bar{P}(m_s) + B(m_s)\bar{K}(m_s)]_s) & B_w(m_s) \\ * & -I \end{bmatrix} \\ = & \sum_{j=1}^{2} \sum_{j=1}^{2} \alpha_i \alpha_j X_{ij} = \sum_{j=1}^{2} \alpha_i X_i + \alpha_1 \alpha_2 \mathrm{X}_{12} \end{aligned} \tag{2.87}$$

where

$$\mathrm{X}_{ij} \triangleq \begin{bmatrix} \bar{P}_i A_j^T + \bar{K}_i B_j^T + A_j \bar{P}_i + B_j \bar{K}_i & B_{wj} \\ * & -I \end{bmatrix}.$$

Therefore, (2.78) holds if

$$\begin{bmatrix} \bar{P}_i A_i^T + \bar{K}_i B_i^T + A_i \bar{P}_i + B_i \bar{K}_i - \tilde{A}_{ii} & B_{wi} - \tilde{B}_{ii} \\ * & -I - \tilde{D}_{ii} \end{bmatrix} < 0, i = 1, 2. \tag{2.88}$$

$$\begin{bmatrix} \mathrm{sym}(A_2 \bar{P}_1 + B_2 \bar{K}_1 + A_1 \bar{P}_2 + B_1 \bar{K}_2 - \tilde{A}_{12}) & B_{w2} + B_{w1} - \tilde{B}_{12} - \tilde{C}_{12}^T \\ * & -2I - \tilde{D}_{12} - \tilde{D}_{12}^T \end{bmatrix} \leq 0,$$

$$\begin{bmatrix} \begin{bmatrix} \tilde{A}_{11} & \tilde{B}_{11} \\ * & \tilde{D}_{11} \end{bmatrix} & \begin{bmatrix} \tilde{A}_{12} & \tilde{B}_{12} \\ \tilde{C}_{12} & \tilde{D}_{12} \end{bmatrix} \\ * & \begin{bmatrix} \tilde{A}_{22} & \tilde{B}_{22} \\ * & \tilde{D}_{22} \end{bmatrix} \end{bmatrix} < 0. \tag{2.89}$$

By using similar techniques, it can be established that (2.79) holds if

$$\begin{bmatrix} -S_i - \tilde{E}_{ii} & C_{1i} \bar{P}_i + D_{1i} \bar{K}_i - \tilde{F}_{ii} \\ * & -\bar{P}_i - \tilde{H}_{ii} \end{bmatrix} < 0, \; i = 1, 2, \tag{2.90}$$

$$\begin{bmatrix} -S_1 - S_2 - \tilde{E}_{12} - \tilde{E}_{12}^T & C_{12} \bar{P}_1 + D_{12} \bar{K}_1 + C_{11} \bar{P}_2 + D_{11} \bar{K}_2 - \tilde{F}_{12} - \tilde{G}_{12}^T \\ * & -\bar{P}_1 - \bar{P}_2 - \tilde{H}_{12} - \tilde{H}_{12}^T \end{bmatrix} \leq 0,$$

$$\begin{bmatrix} \begin{bmatrix} \tilde{E}_{11} & \tilde{F}_{11} \\ * & \tilde{H}_{11} \end{bmatrix} & \begin{bmatrix} \tilde{E}_{12} & \tilde{F}_{12} \\ \tilde{G}_{12} & \tilde{H}_{12} \end{bmatrix} \\ * & \begin{bmatrix} \tilde{E}_{22} & \tilde{F}_{22} \\ * & \tilde{H}_{22} \end{bmatrix} \end{bmatrix} < 0. \tag{2.91}$$

Equation (2.80) holds if

$$\begin{bmatrix} -\gamma_l^2 I - \tilde{I}_{lii} & C_{1i}\bar{P}_i + D_{1i}\bar{K}_i - \tilde{J}_{lii} \\ * & -\bar{P}_i - \tilde{L}_{lii} \end{bmatrix} < 0, \ i = 1, 2, \ l = 2, 3, 4 \tag{2.92}$$

$$\begin{bmatrix} -2\gamma_l^2 I - \tilde{I}_{l12} - \tilde{I}_{l12}^T & C_{l2}\bar{P}_1 + D_{l2}\bar{K}_1 + C_{l1}\bar{P}_2 + D_{l1}\bar{K}_2 - \tilde{J}_{l12} - \tilde{K}_{l12}^T \\ * & -\bar{P}_1 - \bar{P}_2 - \tilde{L}_{l12} - \tilde{L}_{l12}^T \end{bmatrix} \le 0, \tag{2.93}$$
$l = 2, 3, 4,$

$$\begin{bmatrix} \begin{bmatrix} \tilde{I}_{l11} & \tilde{J}_{l11} \\ * & \tilde{L}_{l11} \end{bmatrix} & \begin{bmatrix} \tilde{I}_{l12} & \tilde{J}_{l12} \\ \tilde{K}_{l12} & \tilde{L}_{l12} \end{bmatrix} \\ * & \begin{bmatrix} \tilde{I}_{l12} & \tilde{J}_{l22} \\ * & \tilde{L}_{l22} \end{bmatrix} \end{bmatrix} < 0, l = 2, 3, 4. \tag{2.94}$$

Equation (2.81) holds if

$$\begin{bmatrix} -P_i - \tilde{M}_{ii} & (A_i - \eta I)\bar{P}_i + B_i\bar{K}_i - \tilde{N}_{ii} \\ * & -\rho^2\bar{P}_i - \tilde{P}_{ii} \end{bmatrix} < 0, \ i = 1, 2, \tag{2.95}$$

$$\begin{bmatrix} -\bar{P}_1 - \bar{P}_2 - \tilde{M}_{12} - \tilde{M}_{12}^T & (A_2 - \eta I)\bar{P}_1 + (A_1 - \eta I)\bar{P}_2 + B_1\bar{K}_2 - \tilde{N}_{12} - \tilde{O}_{12}^T \\ * & -\rho^2\bar{P}_1 - \rho^2\bar{P}_2 - \tilde{P}_{12} - \tilde{P}_{12}^T \end{bmatrix} \le 0,$$

$$\begin{bmatrix} \begin{bmatrix} \tilde{M}_{11} & \tilde{N}_{11} \\ * & \tilde{P}_{11} \end{bmatrix} & \begin{bmatrix} \tilde{M}_{12} & \tilde{N}_{12} \\ \tilde{O}_{12} & \tilde{P}_{12} \end{bmatrix} \\ * & \begin{bmatrix} \tilde{M}_{22} & \tilde{N}_{22} \\ * & \tilde{P}_{22} \end{bmatrix} \end{bmatrix} < 0. \tag{2.96}$$

$$\bar{P}_i(A_i - \mu I)^T + (A_i - \mu I)\bar{P}_i + \bar{K}_i^T B_i^T + B_i\bar{K}_i - \tilde{Q}_{ii} < 0, \ i = 1, 2, \tag{2.97}$$

$$\text{sym}((A_2 - \mu I)\bar{P}_1 + B_2\bar{K}_1 + (A_1 - \upsilon I)\bar{P}_2 + B_1\bar{K}_2 - \tilde{Q}_{12}) \le 0, \tag{2.98}$$

$$\begin{bmatrix} \tilde{Q}_{11} & \tilde{Q}_{12} \\ * & \tilde{Q}_{22} \end{bmatrix} < 0. \tag{2.99}$$

Equation (2.83) holds if

$$-\bar{P}_i(A_i - \upsilon I)^T - (A_i - \upsilon I)\bar{P}_i - \bar{K}_i^T B_i^T + B_i\bar{K}_i - \tilde{R}_{ii} < 0, \ i = 1, 2, \tag{2.100}$$

$$\text{sym}(-(A_2 - \upsilon I)\bar{P}_1 - B_2\bar{K}_1 - (A_1 - \upsilon I)\bar{P}_2 - B_1\bar{K}_2 - \tilde{R}_{12}) \le 0, \tag{2.101}$$

$$\begin{bmatrix} \tilde{R}_{11} & \tilde{R}_{12} \\ * & \tilde{R}_{22} \end{bmatrix} < 0. \tag{2.102}$$

Now, we have transformed conditions (2.69), (2.78)–(2.83) into a set of LMI conditions. Based on these conditions, the multi-objective load-dependent controller design in problem load-dependent suspension control (LDSC) can be solved via the following convex optimization problem:

$$\min \gamma_1 \text{ s.t.}(39) - (48) \text{ and OC}, \tag{2.103}$$

where OC refers to the pole placement constraints (2.95)–(2.96) (disk region) or (2.97)–(2.102) (vertical strip region). If the optimization problem (2.103) has a set of feasible solutions, by substituting the matrix functions (2.85) into (2.84), the feedback gain matrix function for controller (2.65) can be given by

$$K(m_s) = \left(\sum_{i=1}^{2} \alpha_i \bar{K}_i\right)\left(\sum_{i=1}^{2} \alpha_i \bar{P}_i\right)^{-1}. \tag{2.104}$$

Remark 2.4 The obtained controller gain matrix function in (2.104) based on the convex optimization problem (2.103) is nonlinearly dependent on the vector α (consequently nonlinearly dependent on m_s), which constitutes the essential difference from previous constant gain controller design.

Remark 2.5 As can be seen in the above derivation process, the Lyapunov matrices for any given body mass ms can be given by

$$P(m_s) = \left(\sum_{i=1}^{2} \alpha_i \bar{P}_i\right)^{-1}$$

which is also dependent on the parameter m_s.

2.2.3 Simulation Verification

Here, we use an example to illustrate the usefulness and advantage of the load-dependent controller design method proposed above. Model parameters are borrowed from [9] and listed in Table 2.1. The values listed in Table 2.1 are for the nominal system. We assume that the sprung mass m_s changes with the vehicle load, which is expressed as

$$m_s = (320 + \lambda)\text{kg},$$

where λ is a parameter satisfying $|\lambda| \leq \bar{\lambda}$. In this case, the state-space model (2.60) can be represented by a two-vertex polytope.

First, assume that $\bar{\lambda} = 64$ kg (that is, the sprung mass ms fluctuates around its nominal value by 20%. In addition, assume the maximum allowable suspension

stroke $z_{\max} = 0.08$ m, the maximum force output $u_{\max} = 1000$ N, the road roughness coefficient $G_0 = 512 \times 10^{-6}\,\mathrm{m}^3$, the reference spatial frequency $q_0 = 0.1\,\mathrm{m}^{-1}$ and the vehicle forward speed $V = 30\,\mathrm{m/s}$. Our purpose is to design a load-dependent controller in the form of (2.65), such that the closed-loop system (2.66) satisfies
(1) $\|T_1(s)\|_2 \le \gamma_1$,
(2) $\|T_l(s)\|_G \le 1,\ l = 2, 3, 4$
(3) All the eigenvalues of $\bar{A}(\lambda)$ lie in the region $\wp(-38, -2)$.

By solving the convex optimization problem (2.103) in the MATLAB environment [10], we have $\gamma_1^* = \min \|T_1(s)\|_2 = 2.7256\,\mathrm{m/s}^2$, and the associated matrices are as follows (for brevity, here we only list the matrices that are necessary for the construction of the admissible controllers):

$$\bar{P}_1 = \begin{bmatrix} 0.0030 & -0.0002 & -0.0095 & -0.0034 \\ -0.0002 & 0.0002 & 0.0006 & -0.0030 \\ -0.0095 & 0.0002 & 0.0311 & 0.0038 \\ -0.0034 & -0.0030 & 0.0038 & 1.0027 \end{bmatrix},$$

$$\bar{P}_2 = \begin{bmatrix} 0.0046 & -0.0001 & -0.00141 & 0.0026 \\ -0.0001 & 0.0002 & 0.0008 & -0.0032 \\ -0.0141 & 0.0008 & 0.1038 & 0.0027 \\ -0.0026 & -0.0032 & 0.0027 & 0.9664 \end{bmatrix},$$

$$\bar{K}_1 = \begin{bmatrix} 51.2114 & -1.5133 & -160.3566 & -393.0366 \end{bmatrix},$$

$$\bar{K}_2 = \begin{bmatrix} 58.1277 & -0.6991 & -243.2434 & -263.5746 \end{bmatrix}.$$

Therefore, the gain matrix function for an admissible load-dependent controller is given by

$$K(m_s) = \left(\sum_{i=1}^{2} \alpha_i \bar{K}_i\right)\left(\sum_{i=1}^{2} \alpha_i \bar{P}_i\right)^{-1}. \tag{2.105}$$

where

$$\begin{aligned} \alpha_1 &= \left(\tfrac{1}{m_s} - \tfrac{1}{320+\bar{\lambda}}\right)/\left(\tfrac{1}{320-\bar{\lambda}} - \tfrac{1}{320+\bar{\lambda}}\right), \\ \alpha_2 &= \left(\tfrac{1}{320-\bar{\lambda}} - \tfrac{1}{m_s}\right)/\left(\tfrac{1}{320-\bar{\lambda}} - \tfrac{1}{320+\bar{\lambda}}\right). \end{aligned} \tag{2.106}$$

Figure 2.9 depicts the eigenvalues of the open- and closed-loop systems in the complex plane, from which we can see that the designed controller renders the poles of the closed-loop system to lie inside the expected region. The H_2 norms of the transfer function $T_1(s)$ for different λ in the admissible interval $|\lambda| \le \bar{\lambda}$ are shown in Fig. 2.10. It can be seen from this figure that for all admissible parameter λ, we have $\|T_1(s)\|_2 \le \gamma^* = 2.7256\,\mathrm{m/s}^2$. In addition, $\|T_l(s)\|_G, l = 2, 3, 4$ for different admissible λ are also presented in Figs. 2.11, 2.12 and 2.13, which clearly show $\|T_l(s)\|_G < 1$.

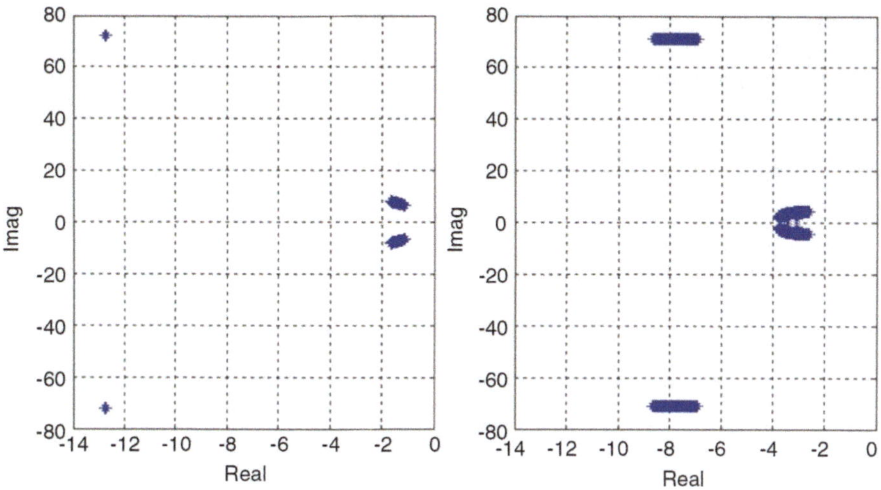

Fig. 2.9 Poles of open- and closed-loop systems

Fig. 2.10 $\|T_1(s)\|_2$ of open- and closed-loop systems versus parameter λ

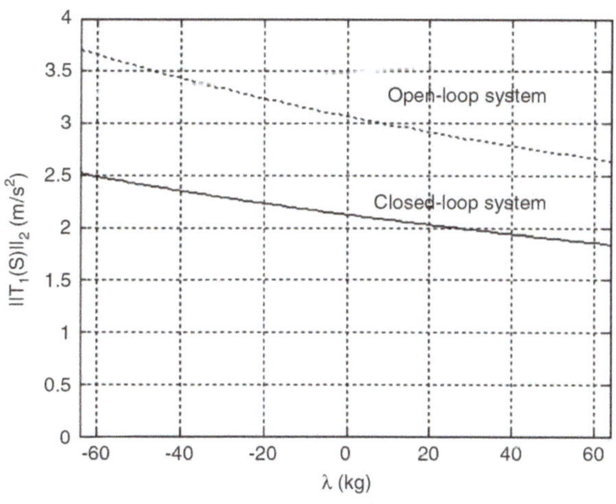

As is mentioned above, the controller gain matrix function (2.105) is in fact a nonlinear function in terms of the parameter λ. In order to see clear the relationship between $K(m_s)$ and λ, Fig. 2.14 depicts the four components of $K(m_s)$ for different λ.

Figure 2.15 shows the open- and closed-loop frequency responses from the ground vertical velocity $\dot{z}_r$ to the body acceleration $\ddot{z}_s$. From this figure we can see that the closed-loop system has a significant reduction in amplitude when compared with the open-loop system, especially for the frequency band (4–8 Hz), in which the human body is more sensitive to vertical vibration. Therefore, the ride comfort has been improved significantly under the designed load-dependent controller.

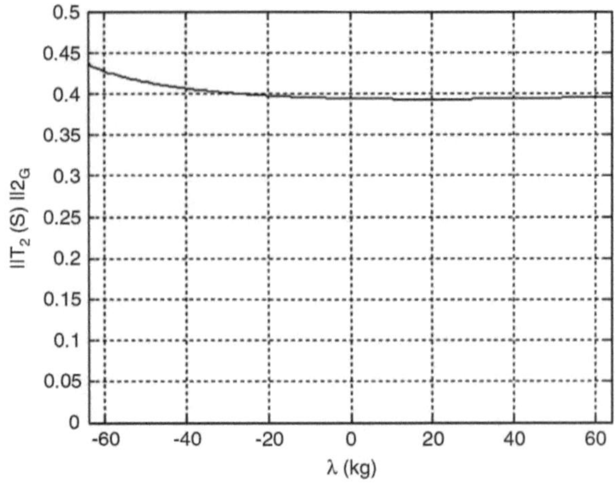

Fig. 2.11 $\|T_2(s)\|_G$ of open- and closed-loop systems versus parameter λ

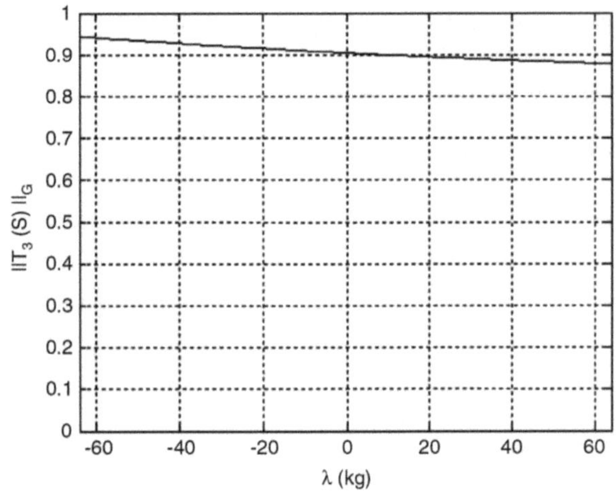

Fig. 2.12 $\|T_3(s)\|_G$ of open- and closed-loop systems versus parameter λ

Now assume the disturbance input from the ground $w(t)$ to be zero-mean white noise with identity power spectral density (shown in Fig. 2.16). Figure 2.17 shows the body accelerations of the open- and closed-loop systems, from which we can see the effectiveness of the designed load-dependent controller (in this figure, the solid line, dashed line and dotted line represent the case $\lambda = -64$, $\lambda = 0$ and $\lambda = 64$, respectively).

Finally, a comparison between the load-dependent controller design and the constant controller design presented in [5] is carried out. Table 2.3 lists the obtained minimum H_2 norm of $T_1(s)$ for different $\bar{\lambda}$. It can be seen that under the same conditions, the load-dependent controller approach can yield much less conservative designs than the constant gain approach. Notably for $\bar{\lambda} = 144$ where the constant controller method fails to find feasible solutions, our load-dependent approach is still able to provide desired controllers. To highlight the benefit of the load-dependent

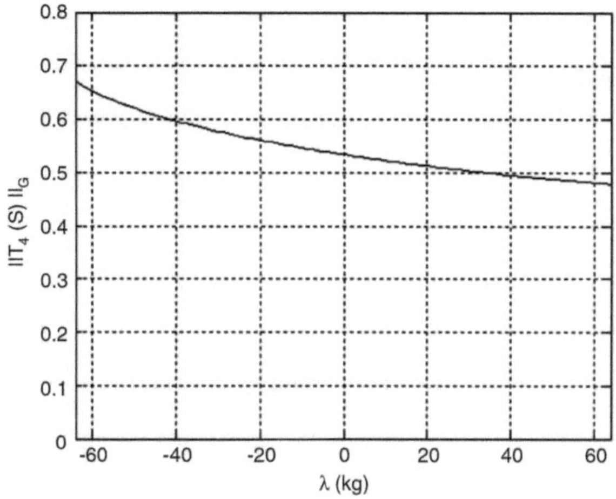

Fig. 2.13 $\|T_4(s)\|_G$ of open- and closed-loop systems versus parameter λ

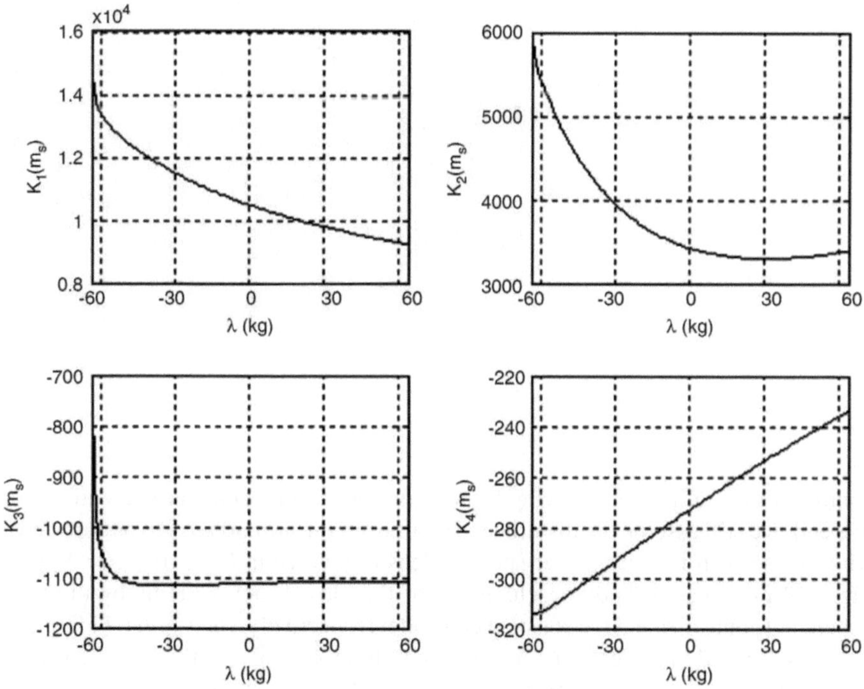

Fig. 2.14 Nonlinear behavior of controller gains versus parameter λ

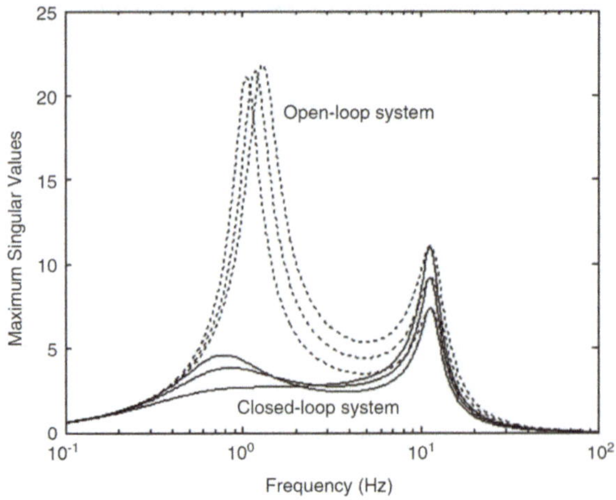

Fig. 2.15 Frequency response of open- and closed-loop systems from ground velocity $\dot{z}_r$ to body acceleration $\ddot{z}_s$ ($\lambda = -64, 0, 64$)

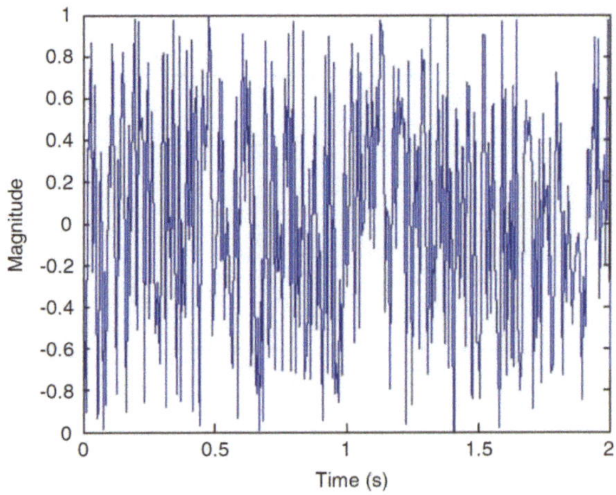

Fig. 2.16 Disturbance input from the ground $w(t)$

controller design, in the following we will present some computer simulations. To this end, we still assume the disturbance input from the ground $w(t)$ to be zero-mean white noise with identity power spectral density (shown in Fig. 2.16). For $\bar{\lambda} = 32$, Fig. 2.18 presents the body accelerations of the closed-loop systems by the load-dependent controller and the constant controller respectively; for $\bar{\lambda} = 64$. Figure 2.19 presents the body accelerations of the closed-loop systems by the load-dependent controller and the constant controller, respectively. From these figures, we can see that the load-dependent controller design yields better controllers than the constant controller design approach.

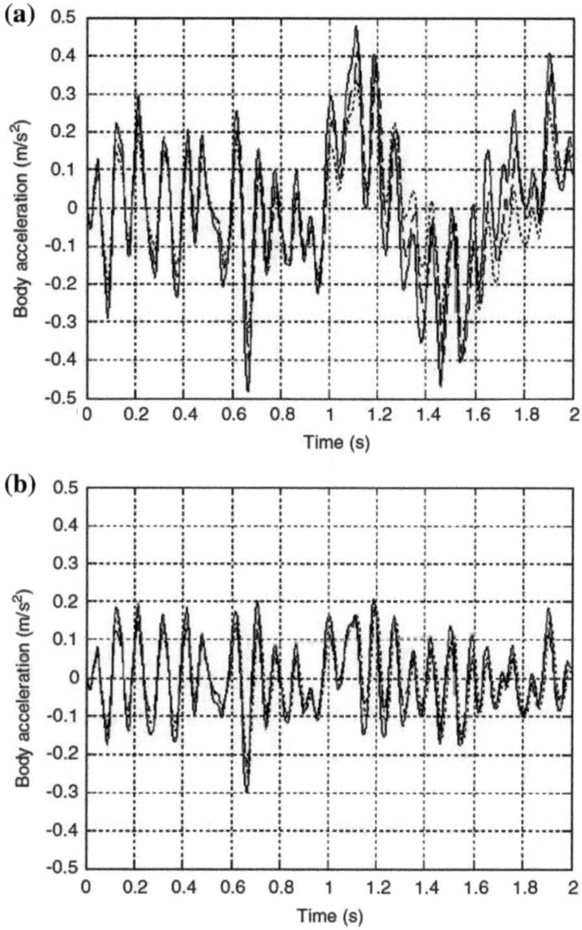

Fig. 2.17 Vertical accelerations of **a** open- and **b** closed-loop systems ($\lambda = -64, 0, 64$)

Table 2.3 Obtained minimum $\|T_1(s)\|_2$ (comparing results with [4])

$\bar{\lambda}$(kg)	32	64	96	128	144
Our method m/s^2	2.2608	2.7256	3.4386	4.5863	5.4659
Chen et al. [4] m/s^2	2.5739	3.2689	4.2967	5.9526	Infeasible

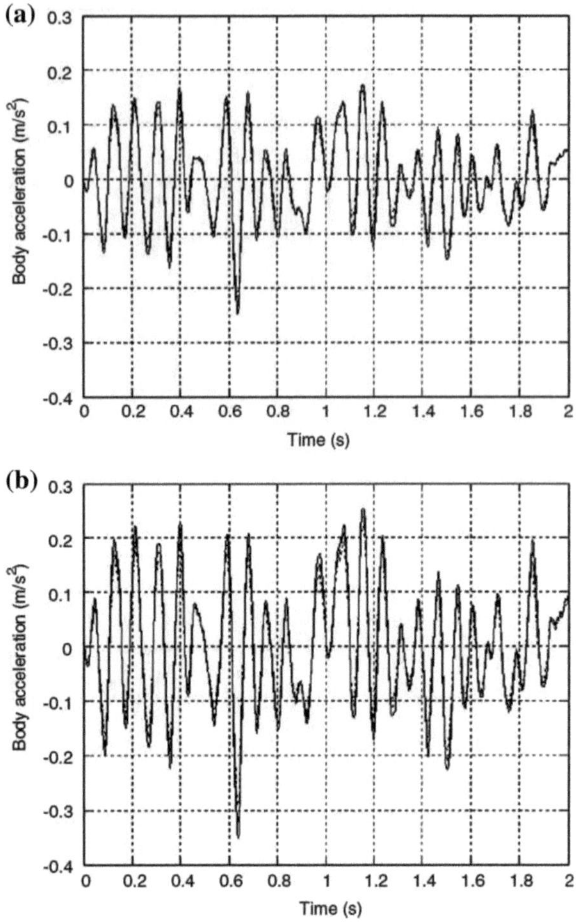

Fig. 2.18 Vertical accelerations of closed-loop systems by different controllers for $\bar{\lambda} = 32$: **a** load-dependent Controller; **b** Constant controller ($\lambda = -32, 0, 32$)

2.2.4 *Conclusion*

A load-dependent controller design approach has been proposed to solve the problem of multiobjective control of active suspension systems with uncertain parameters. This approach designs controllers whose gain matrix depends on the online available information of the body mass based on parameter-dependent Lyapunov functions. Compared with previous approaches that design robust constant controllers, the proposed load-dependent approach can yield much less conservative results. The usefulness and the advantages of the proposed controller design methodology are illustrated via simulations. Finally, it is worth mentioning that as only statefeedback case is considered in this paper, future research effort can be directed at solving the problem of output-feedback controller design (such as that considered in [11]), which is more suitable for the case when some of the state variables are not measurable.

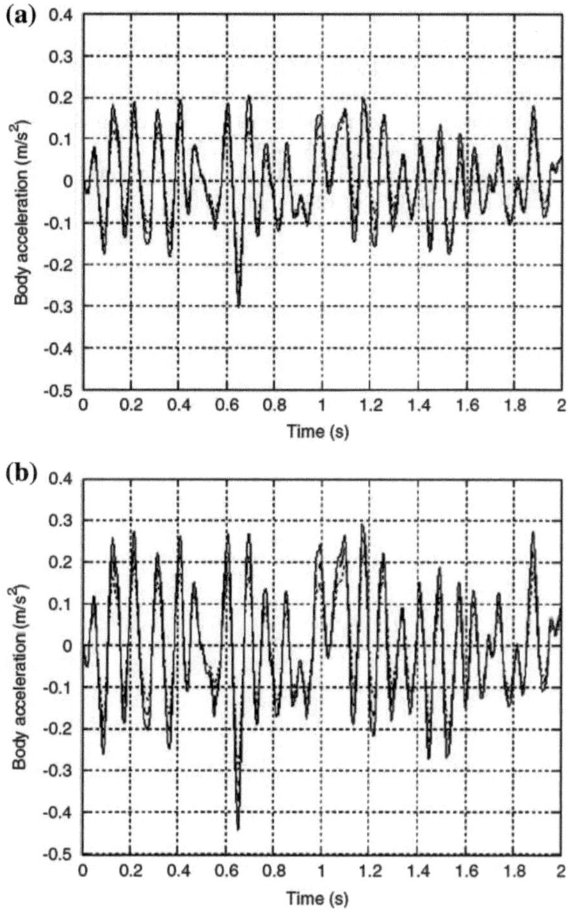

Fig. 2.19 Vertical accelerations of closed-loop systems by different controllers for $\bar{\lambda} = 64$: **a** load-dependent controller; **b** constant controller ($\lambda = -64, 0, 64$)

References

1. H. Chen, K.H. Guo, Constrained H_∞ control of active suspensions: an LMI approach. IEEE Trans. Control. Syst. Technol. **13**(3), 412–421 (2005)
2. S. Boyd, L.E. Ghaoui, E. Feron, V. Balakishnan, *Linear Matrix Inequalities in System and Control Theory* (SIAM, Philadelphia, PA, 1994)
3. D. Hrovat, Survey of advanced suspension developments and related optimal control applications. Automatica **33**(10), 1781–1817 (1997)
4. H. Chen, P.Y. Sun, K.H. Guo, A multi-objective control design for active suspensions with hard constraints. In: Proceedings of the American Control Conference (2003), pp. 14371–4376
5. H. Gao, C. Wang, Robust energy-to-peak filtering with improved LMI representations. IEE Proc. Vis. Image Signal Process. **150**, 82–89 (2003)
6. H. Gao, C. Wang, A delay-dependent approach to robust HN and L2CLN filtering for a class of uncertain nonlinear time-delayed systems. IEEE Trans. Autom. Control. **48**(9), 1661–1666 (2003)
7. M.A. Rotea, The generalized H_2 control problem. Automatica **29**, 73C385 (1993)

8. M. Chilali, P. Gahinet, H_∞ design with pole placement constraints: an LMI approach. IEEE Trans. Autom. Control. **41**(3), 358–367 (1996)
9. T.J. Gordon, C. Marsh, M.G. Milsted, A comparison of adaptive LQG and nonlinear controllers for vehicle suspension systems. Veh. Syst. Dyn. **20**(6), 321–340 (1991)
10. P. Gahinet, A. Nemirovskii, A.J. Laub, M. Chilali, *LMI Control Toolbox Users Guide* (The Math. Works Inc., Natick, MA, 1995)
11. M.M. El Madany, M.I. Al-Majed, Quadratic synthesis of active controls for a quarter-car model. J. Vib. Control. **7**(8), 1237–1252 (2001)

Chapter 3
Finite Frequency H_∞ Control for Active Suspensions

Compared with the H_∞ control method in the entire frequency domain in Sect. 2.1, a H_∞ control scheme in the finite frequency domain is proposed in this chapter. In Sect. 3.1, the finite frequency H_∞ control is more effective because the human body is much sensitive to vibrations of a certain frequency band. The H_∞ control scheme in finite frequency domain is based on that all the state variables are observable in Sect. 3.1. However, in practice, it is difficult to realize all the state variables measurable. And obtaining all the state variables resulted in more sensors, higher cost, and additional complexity. Therefore, in Sect. 3.2, the dynamic output feedback control is suggested in the finite frequency domain in consideration of not all the state variables being measurable.

Notation: For a matrix P, P^{T}, P^*, P^{-1} and $P^{\perp}$ denote its transpose, conjugate transpose, inverse and orthogonal complement, respectively; the notation $P > 0$ (≥ 0) means that P is real symmetric and positive definite (semi-definite); and $[P]_s$ means $P + P^{\mathrm{T}}$. $\|G\|_\infty$ denotes the H_∞-norm of transfer function matrix $G(s)$. For matrices P and Q, $P \otimes Q$ means the Kronecker product. In symmetric block matrices or complex matrix expressions, we use an asterisk ($*$) to represent a term that is induced by symmetry and diag$\{\ldots\}$ stands for a block-diagonal matrix. Matrices, if their dimensions are not explicitly stated, are assumed to be compatible for algebraic operations. The space of square-integrable vector functions over $[0, \infty)$ is denoted by $L_2[0, \infty)$, and for $w = \{w(t)\} \in L_2[0, \infty)$, its norm is given by $\|w\|_2 = \sqrt{\int_{t=0}^{\infty} |w(t)|^2 dt}$.

W. Sun et al., *Advanced Control for Vehicle Active Suspension Systems*, Studies in Systems, Decision and Control 204,
https://doi.org/10.1007/978-3-030-15785-2_3

3.1 Static State Feedback H_∞ Control for Active Suspensions

This section addresses the problem of H_∞ control for active vehicle suspension systems in finite frequency domain. The H_∞ performance is used to measure ride comfort so that more general road disturbances can be considered. By using the generalized Kalman–Yakubovich–Popov (KYP) lemma, the H_∞ norm from the disturbance to the controlled output is decreased in a specific frequency band to improve the ride comfort. Compared with the entire frequency approach, the finite frequency approach suppresses the vibration more effectively for the concerned frequency range. In addition, the time-domain constraints, which represent performance requirements for vehicle suspensions, are guaranteed in the controller design. A state feedback controller is designed in the framework of LMI optimization. A quarter-car model with active suspension system is considered in this section and a numerical example is employed to illustrate the effectiveness of the proposed approach.

3.1.1 Finite Frequency H_∞ Control Scheme

The most important objective for vehicle suspension systems is the improvement of ride comfort. In other words, the main task is to design the controller which can succeed in stabilizing the vertical motion of the car body and isolating the force transmitted to the passengers as well. It is worth mentioning that most of the reported approaches are considered in the entire frequency domain. However, active suspension systems may just belong to a certain frequency band, and ride comfort is known to be frequency sensitive. From the ISO2631, the human body is much sensitive to vibrations of 4–8 Hz in the vertical direction. Hence, the development of H_∞ control in finite frequency domain is significative for active suspension systems.

The traditional approach for finite frequency domain is to introduce the weighting functions. The weighting method is useful in practice, however, the additional weights increase the system complexity. Besides, the process of selecting appropriate weights is time-consuming, especially when the designer has to shoot for a good trade-off between the complexity of the weights and the accuracy in capturing desired specifications. An alternative approach is to grid the frequency axis. This approach has a practical significance especially when the system is well damped and the frequency response is expected to be smooth. But it lacks a rigorous performance guarantee in the design process [1].

Another approach that avoids both weighting functions and frequency gridding is to generalize the fundamental machinery, the Kalman–Yakubovič–Popov (KYP) lemma. The KYP lemma establishes the equivalence between a frequency domain inequality for a transfer function and a linear matrix inequality (LMI) associated with its state-space realization [2–4]. It allows us to characterize various properties of dynamic systems in the frequency domain in terms of LMIs. However, the standard

KYP lemma is only applicable for the infinite frequency range. Recently, a very significant development made by Iwasaki and Hara is the generalized KYP lemma [5]. It establishes the equivalence between a frequency domain property and an LMI over a finite frequency range, allowing designers to impose performance requirements over chosen finite or infinite frequency ranges. The generalized KYP lemma is very useful for the analysis and synthesis problems in practical applications.

Different from the conventional methodologies that consider the H_∞ control over the entire frequency range, we consider the active suspension systems over the finite frequency range based on the generalized KYP lemma. In addition, the time-domain constraints (road holding, suspension deflection and actuator saturation) are guaranteed in the controller design. By using the generalized KYP lemma, the frequency domain inequalities are transformed into linear matrix inequalities, and our attention is focused on developing methods to design a state feedback control law based on matrix inequalities such that the resulting closed-loop system is asymptotically stable with a prescribed level of disturbance attenuation in certain frequency domain.

Problem Formulation

The quarter-car model described in Fig. 1.2 is considered in this section. Also, the effect of actuator dynamics is neglected and the actuator is modelled as an ideal force generator. The state-space expression of the quarter-car model is derived as (2.8) and (2.9) in Sect. 2.1.

It is widely accepted that ride comfort is closely related to the body acceleration in frequency band 4–8 Hz. Consequently, in order to improve ride comfort, it is important to keep the transfer function from the disturbance inputs, $w(t)$, to car body acceleration, $\ddot{z}_s(t)$, as small as possible over the frequency band 4–8 Hz.

Denote $G(j\omega)$ as the transfer function from the disturbance inputs $w(t)$ to the controlled output $z_1(t)$. The finite frequency H_∞ control problem is to design a controller such that the closed-loop system guarantees

$$\sup_{\varpi_1<\omega<\varpi_2} \|G(j\omega)\|_\infty < \gamma, \tag{3.1}$$

where $\gamma > 0$ is a prescribed scalar, and ϖ_1, ϖ_2 represent the lower and upper bounds of the concerned frequency. In addition, from the safety and mechanical structure point of view, the constraints (2.10) need to be guaranteed.

Controller Design

To facilitate the presentation, essential lemmas are introduced. For the sake of brevity, all the proof of the lemmas have been omitted.

Lemma 3.1 (Generalized KYP Lemma [5]) *Consider the linear system* $(\bar{A}, \bar{B}, \bar{C}, \bar{D})$. *Given a symmetric matrix* Π , *the following statements are equivalent:*

(1) The finite frequency inequality

$$\begin{bmatrix} G(j\omega) \\ I \end{bmatrix}^T \Pi \begin{bmatrix} G(j\omega) \\ I \end{bmatrix} < 0, \ \varpi_1 \le \omega \le \varpi_2. \tag{3.2}$$

(2) There exist symmetric matrices P and Q satisfying $Q > 0$ and

$$\begin{bmatrix} \Gamma[P, Q, \bar{C}, \bar{D}] & [\bar{C}\ \bar{D}]^T \\ * & -I \end{bmatrix} < 0, \tag{3.3}$$

where

$$\Gamma[P, Q, \bar{C}, \bar{D}] = \begin{bmatrix} \bar{A} & \bar{B} \\ I & 0 \end{bmatrix}^T \begin{bmatrix} -Q & P + j\varpi_c Q \\ P - j\varpi_c Q & -\varpi_1\varpi_2 Q \end{bmatrix} \begin{bmatrix} \bar{A} & \bar{B} \\ I & 0 \end{bmatrix} + \begin{bmatrix} 0 & \bar{C}^T \Pi_{12} \\ * & [\bar{D}^T \Pi_{12}]_s + \Pi_{22} \end{bmatrix}, \tag{3.4}$$

$\varpi_c = (\varpi_1 + \varpi_2)/2$, and Π_{12}, Π_{22} are the upper right and lower right block matrices of Π.

Lemma 3.2 (Projection Lemma [6]) *Let Γ, Λ, Θ be given, there exists a matrix F satisfying $\Gamma F \Lambda + (\Gamma F \Lambda)^T + \Theta < 0$ if and only if the two conditions hold: $\Gamma^{\perp} \Theta \Gamma^{\perp T} < 0$, $\Lambda^{T^{\perp}} \Theta \Lambda^{T^{\perp T}} < 0$.*

Lemma 3.3 (Reciprocal Projection Lemma [6]) *Let P be any given positive definite matrix. The inequality $\Psi + S + S^T < 0$ is equivalent to the LMI problem:*

$$\begin{bmatrix} \Psi + P - [W]_s & S^T + W^T \\ * & -P \end{bmatrix} < 0. \tag{3.5}$$

It is assumed that all the state variables can be measured, and we are interested in designing a state feedback controller

$$u(t) = Kx(t), \tag{3.6}$$

where K is the state feedback gain matrix to be designed. By combining (3.6) with (2.8), the closed-loop system is given by:

$$\begin{aligned} \dot{x}(t) &= \bar{A}x(t) + \bar{B}w(t), \\ z_1(t) &= \bar{C}x(t) + \bar{D}w(t), \\ z_2(t) &= C_2 x(t), \end{aligned} \tag{3.7}$$

where

$$\left[\begin{array}{c|c} \bar{A} & \bar{B} \\ \hline \bar{C} & \bar{D} \end{array}\right] = \left[\begin{array}{c|c} A + BK & B_w \\ \hline C_1 + D_1 K & 0 \end{array}\right]. \tag{3.8}$$

For the active suspension systems, in accordance with the requirements, the constrained H_∞ control problem is formulated to minimize the H_∞ norm from the disturbance inputs $w(t)$ to the controlled output $z_1(t)$ under the time-domain constraints (2.10) over the fixed frequency band $\varpi_1 \le \omega \le \varpi_2$. By using Lemma 2.1, we have the following theorem.

Theorem 3.4 *Let positive scalars γ, η and ρ be given. A state feedback controller in the form of (3.6) exists, such that the closed-loop system in (3.7) is asymptotically stable with $w(t) = 0$, and satisfies $\|G(j\omega)\|_\infty^{\varpi_1<\omega<\varpi_2} < \gamma$ for all nonzero $w \in L_2[0, \infty)$, while the constraints in (2.10) are guaranteed with the disturbance energy under the bound $w_{\max} = (\rho - V(0))/\eta$, if there exist symmetric matrices P, $P_1 > 0$, $Q > 0$ and general matrix F satisfying*

$$\begin{bmatrix} -[F]_s & F^T\bar{A} + P_1 & F^T & F^T\bar{B} \\ * & -P_1 & 0 & 0 \\ * & * & -P_1 & 0 \\ * & * & * & -\eta I \end{bmatrix} < 0, \tag{3.9}$$

$$\begin{bmatrix} -Q & P + j\varpi_c Q - F & 0 & 0 \\ * & -\varpi_1\varpi_2 Q + [F^T\bar{A}]_s & F^T\bar{B} & \bar{C}^T \\ * & * & -\gamma^2 I & 0 \\ * & * & * & -I \end{bmatrix} < 0, \tag{3.10}$$

$$\begin{bmatrix} I & \sqrt{\rho}K \\ * & -u_{\max}^2 P_1 \end{bmatrix} \le 0, \tag{3.11}$$

$$\begin{bmatrix} -I & \sqrt{\rho}\{C_2\}_i \\ * & -P_1 \end{bmatrix} < 0, \ i = 1, 2, \tag{3.12}$$

where $\varpi_c = (\varpi_1 + \varpi_2)/2$ is a given scalar.

Proof By using Schur complement, inequality (3.9) is equivalent to

$$\begin{bmatrix} \frac{1}{\eta}F^T\bar{B}\bar{B}^T F + F^T P_1^{-1} F - [F]_s & F^T\bar{A} + P_1 \\ * & -P_1 \end{bmatrix} < 0. \tag{3.13}$$

Performing the congruence transformation to inequality (3.13) by diag $\{F^{-1}, P_1^{-1}\}$, with $F := W^{-1}$, inequality (3.13) can be transformed to the following inequality:

$$\begin{bmatrix} \frac{1}{\eta}\bar{B}\bar{B}^T + P_1^{-1} - [W]_s & \bar{A}P_1^{-1} + W^T \\ * & -P_1^{-1} \end{bmatrix} < 0. \tag{3.14}$$

By using Lemma 3.3, inequality (3.14) is equivalent to $\bar{A}P_1^{-1} + P_1^{-1}\bar{A}^T + \frac{1}{\eta}\bar{B}\bar{B}^T < 0$, with $\Psi = \frac{1}{\eta}\bar{B}\bar{B}^T$ and $S^T = \bar{A}P_1^{-1}$. Clearly, we have

$$\bar{A}^T P_1 + P_1\bar{A} + \frac{1}{\eta}P_1\bar{B}\bar{B}^T P_1 < 0, \tag{3.15}$$

which can guarantee $\bar{A}^T P_1 + P_1 \bar{A} < 0$. From the standard Lyapunov theory for continuous-time linear system, the closed-loop system (3.7) is asymptotically stable with $w(t) = 0$.

Rewrite inequality (3.10) as

$$J \Xi J^T + H \Pi H^T + [\Gamma F \Lambda]_s < 0, \tag{3.16}$$

where

$$J = \begin{bmatrix} I & 0 & 0 \\ 0 & I & 0 \end{bmatrix}^T, \ \Xi = \begin{bmatrix} -Q & P + j\varpi_c Q \\ * & -\varpi_1 \varpi_2 Q \end{bmatrix}, \ \Pi = \begin{bmatrix} I & 0 \\ * & -\gamma^2 I \end{bmatrix}, \tag{3.17}$$

$$H = \begin{bmatrix} 0 & \bar{C} & 0 \\ 0 & 0 & I \end{bmatrix}^T, \ \Gamma = \begin{bmatrix} -I & \bar{A} & \bar{B} \end{bmatrix}^T, \ \Lambda = \begin{bmatrix} 0 & I & 0 \end{bmatrix}. \tag{3.18}$$

Then, according to Lemma 3.2, inequality (3.16) holds if and only if

$$\begin{aligned} W^T (J \Xi J^T + H \Pi H^T) W &< 0, \\ U (J \Xi J^T + H \Pi H^T) U^T &< 0, \end{aligned} \tag{3.19}$$

where

$$W = \begin{bmatrix} I & 0 & 0 \\ 0 & 0 & I \end{bmatrix}^T, \ U = \begin{bmatrix} \bar{A}^T & I & 0 \\ \bar{B}^T & 0 & I \end{bmatrix}.$$

Note that inequality (3.19) can be transformed to the following form:

$$\begin{bmatrix} \bar{A} & \bar{B} \\ I & 0 \end{bmatrix}^T \Xi \begin{bmatrix} \bar{A} & \bar{B} \\ I & 0 \end{bmatrix} + \begin{bmatrix} \bar{C} & 0 \\ 0 & I \end{bmatrix}^T \Pi \begin{bmatrix} \bar{C} & 0 \\ 0 & I \end{bmatrix} < 0, \tag{3.20}$$

which can be further transformed to

$$L + \begin{bmatrix} \bar{C} & 0 \end{bmatrix}^T \begin{bmatrix} \bar{C} & 0 \end{bmatrix} < 0, \tag{3.21}$$

where

$$L = \begin{bmatrix} \bar{A} & \bar{B} \\ I & 0 \end{bmatrix}^T \Xi \begin{bmatrix} \bar{A} & \bar{B} \\ I & 0 \end{bmatrix} + \begin{bmatrix} 0 & 0 \\ 0 & -\gamma^2 I \end{bmatrix}.$$

By using Schur complement and Lemma 3.1, we can obtain

$$\begin{bmatrix} G(j\omega) \\ I \end{bmatrix}^T \Pi \begin{bmatrix} G(j\omega) \\ I \end{bmatrix} < 0, \ \varpi_1 \le \omega \le \varpi_2, \tag{3.22}$$

which is exactly the finite frequency H_∞ performance index inequality in (3.1).

Denote $V(t) = x^T(t)P_1x(t)$ as the energy function, and noting that

$$2x^T(t)P_1\bar{B}w(t) \le \frac{1}{\eta}x(t)^T P_1\bar{B}\bar{B}^T P_1x(t) + \eta w(t)^T w(t), \forall\eta > 0,$$

we have

$$\dot{V}(t) \le x(t)^T(\bar{A}^T P_1 + P_1\bar{A} + \frac{1}{\eta}P_1\bar{B}\bar{B}^T P_1)x(t) + \eta w(t)^T w(t). \tag{3.23}$$

According to the inequality in (3.15), inequality (3.23) guarantees $\dot{V}(t) \le \eta w(t)^T w(t)$. Integrating both sides of the above inequality ($\dot{V}(t) \le \eta w(t)^T w(t)$) from 0 to t results in

$$V(t) - V(0) \le \eta \int_0^t w^T(t)w(t)dt \le \eta \|w\|_2^2 = \eta w_{\max}.$$

This shows that

$$x^T(t)P_1x(t) \le V(0) + \eta w_{\max} = \rho. \tag{3.24}$$

Consider

$$\max_{t\ge0} |u(t)|^2 = \max_{t\ge0} \|Kx(t)\|_2^2 = \max_{t\ge0} \left\|x^T(t)K^T Kx(t)\right\|_2,$$

$$\max_{t\ge0} |\{z_2(t)\}_i|^2 = \max_{t\ge0} \left\|x^T(t)\{C_2\}_i^T\{C_2\}_i x(t)\right\|_2, \; i = 1, 2.$$

Using the transformation $\bar{x}(t) = P_1^{\frac{1}{2}}x(t)$, from inequality (3.24) it follows that $\bar{x}^T(t)\bar{x}(t) \le \rho$. Hence,

$$\max_{t\ge0} |u(t)|^2 = \max_{t\ge0} \left\|\bar{x}^T(t)P_1^{-\frac{1}{2}}K^T KP_1^{-\frac{1}{2}}\bar{x}(t)\right\|_2 \le \rho\cdot\lambda_{\max}\left(P_1^{-\frac{1}{2}}K^T KP_1^{-\frac{1}{2}}\right),$$

$$\max_{t\ge0} |\{z_2(t)\}_i|^2 \le \rho\cdot\lambda_{\max}\left(P_1^{-\frac{1}{2}}\{C_2\}_i^T\{C_2\}_i P_1^{-\frac{1}{2}}\right), \; i = 1, 2, \tag{3.25}$$

where $\lambda_{\max}(\cdot)$ represents the maximum eigenvalue. Then, the constraints in (2.10) hold if

$$\rho P_1^{-\frac{1}{2}}K^T KP_1^{-\frac{1}{2}} < u_{\max}^2 I, \; \rho P_1^{-\frac{1}{2}}\{C_2\}_i^T\{C_2\}_i P_1^{-\frac{1}{2}} < I, \; i = 1, 2, \tag{3.26}$$

which, by Schur complement, are equivalent to (3.11) and (3.12). The proof is completed. □

Since expressions like (3.9) and (3.10) involve the forms of FBK, the resulting feasibility problem is nonlinear. Hence, it cannot be handled directly by LMI optimization. In order to solve the nonlinear problem, define $J_1 = \text{diag}\left\{F^{-1}, F^{-1}, F^{-1}, I\right\}$, $J_2 = \text{diag}\left\{F^{-1}, F^{-1}, I, I\right\}$, $J_3 = \text{diag}\left\{I, F^{-1}\right\}$. Then, we perform a congruence transformation to (3.9), (3.10), (3.11), (3.12), respectively, by the full rank matrix J_1^T, J_2^T, J_3^T and J_3^T on the left, and J_1, J_2, J_3 and J_3 on the right. Defining

$$\begin{aligned}\bar{Q} &= (F^{-1})^T Q F^{-1}, \bar{P} = (F^{-1})^T P F^{-1}, \bar{P}_1 = (F^{-1})^T P_1 F^{-1},\\ \bar{K} &= K F^{-1}, \bar{F} = F^{-1},\end{aligned} \tag{3.27}$$

the following theorem is obtained.

Theorem 3.5 *Let positive scalars γ, η and ρ be given. A state feedback controller in the form of (3.6) exists, such that the closed-loop system in (3.7) is asymptotically stable with $w(t) = 0$, and satisfies $\|G(j\omega)\|_\infty^{\varpi_1<\omega<\varpi_2} < \gamma$ for all nonzero $w \in L_2[0, \infty)$, while the constraints in (2.10) are guaranteed with the disturbance energy under the bound $w_{\max} = (\rho - V(0))/\eta$, if there exist matrices $\bar{P}$, $\bar{P}_1 > 0$, $\bar{Q} > 0$ and general matrix $\bar{F}$ satisfying*

$$\begin{bmatrix} -\left[\bar{F}\right]_s & A\bar{F} + B\bar{K} + \bar{P}_1 & \bar{F} & B_1 \\ * & -\bar{P}_1 & 0 & 0 \\ * & * & -\bar{P}_1 & 0 \\ * & * & * & -\eta I \end{bmatrix} < 0, \tag{3.28}$$

$$\begin{bmatrix} -\bar{Q} & \bar{P} + j\varpi_c\bar{Q} - \bar{F} & 0 & 0 \\ * & -\varpi_1\varpi_2\bar{Q} + [A\bar{F} + B\bar{K}]_s & B_1 & \bar{F}^T C_1^T + \bar{K}^T D_1^T \\ * & * & -\gamma^2 I & 0 \\ * & * & * & -I \end{bmatrix} < 0, \tag{3.29}$$

$$\begin{bmatrix} -I & \sqrt{\rho}\bar{K} \\ * & -u_{\max}^2\bar{P}_1 \end{bmatrix} < 0, \tag{3.30}$$

$$\begin{bmatrix} -I & \sqrt{\rho}\{C_2\}_i\bar{F} \\ * & -\bar{P}_1 \end{bmatrix} < 0, \; i = 1, 2. \tag{3.31}$$

Moreover, if inequalities (3.28)–(3.31) have a set of feasible solutions, the control gain K in (3.6) is given by

$$K = \bar{K}\bar{F}^{-1}.$$

Remark 3.6 Note that the linear matrix inequality (3.29) has complex variables. According to [7], the LMI in complex variables can be converted to an LMI of larger dimension in real variables. This means that inequality $S_1 + jS_2 < 0$ is equivalent to $\begin{bmatrix} S_1 & S_2 \\ -S_2 & S_1 \end{bmatrix} < 0$, which implies the LMI in (3.29) can be addressed.

3.1.2 Simulation Verification

Here, we will apply the above approach to designing a finite frequency state feedback H_∞ controller based on the quarter-car model described in Sect. 2.1. The quarter-car model parameters are the same as those used in entire frequency controller design, listed in Table 2.1.

For subsequent comparison, a state feedback H_∞ controller in the finite frequency domain for system (3.7) is designed first, based on the assumption that all the state variables can be measured. Under zero initial conditions, solve the matrix inequalities (3.28)–(3.31) for matrices $\bar{P}$, $\bar{P}_1 > 0$ and $\bar{Q} > 0$ with the optimized parameter $\gamma > 0$ and maximums of time-domain constraints and other scalar values are listed in Table 3.1.

In the case of optimal γ, an admissible control gain matrix is given based on $K_F = \bar{K}\bar{F}^{-1}$:

$$K_F = 10^4 \times \begin{bmatrix} 0.5033 & -1.3155 & -0.5329 & -0.0547 \end{bmatrix}.$$

For description in brevity, we denote this finite frequency controller as Controller I hereafter.

Then, we give another H_∞ state feedback controller which is designed over the entire frequency range, that is :

$$K_E = 10^4 \times \begin{bmatrix} 1.3900 & 0.4263 & -0.0932 & -0.0400 \end{bmatrix},$$

and we denote this controller as Controller II for brevity.

After obtaining the finite and entire frequency controller, we will compare the two controllers to illustrate the performance of the closed-loop suspension system in finite frequency domain. By the simulation, the responses of the open-loop system, the closed-loop system which is composed of the Controller I and the closed-loop system which is composed of the Controller II, are compared in Fig. 3.1. In this figure, the solid and dotted lines are the responses of the closed-loop system with finite frequency controller and entire frequency controller, respectively, and the dashed line is the response of the passive system. From the figure, we can see that the finite

Table 3.1 Parameters of the suspension system

The maximum suspension deflection $x_{1\,max}$	0.1 m
The maximum tire deflection $x_{2\,max}$	0.01764 m
The maximum active force u_{max}	2500 N
The positive scalar ρ	0.9
The positive scalar η	10000
The lower bound ϖ_1	4 Hz
The upper bound ϖ_2	8 Hz

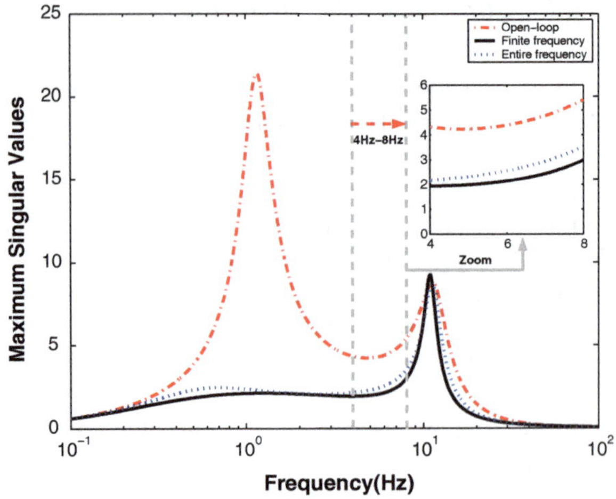

Fig. 3.1 The frequency response of body vertical acceleration

frequency controller yields the least value of H_∞ norm over the frequency range 4–8 Hz, compared with the passive system and the closed-loop system with an entire frequency controller, which clearly shows that an improved ride comfort has been achieved.

In order to evaluate the suspension characteristics with respect to three performance requirements, we give the disturbance signal as follows to clarify the effectiveness of our finite frequency controller.

Consider the case of an isolated bump in an otherwise smooth road surface, the disturbance inputs are given by (1.9), which is equivalent to

$$w(t) = \begin{cases} a\sin(2\pi f t), & \text{if } 0 \le t \le T, \\ 0, & \text{if } t > T, \end{cases} \tag{3.32}$$

where a, f and T represent the amplitude, frequency and period of vibration, respectively. Assume $a = 0.5\,\text{m/s}$, $f = 5\,\text{Hz}$ (among the frequency band 4–8 Hz) and $T = 1/f = 0.2\,\text{s}$. The time-domain response of body vertical acceleration for the active suspension system is shown in Fig. 3.2, where the black solid line and the red dashed line are the responses of body vertical acceleration with the finite frequency controller and the entire frequency controller, respectively. We can clearly see that the value of the body acceleration with the finite frequency controller is less than that with the entire frequency controller. In addition, Fig. 3.3 shows that the ratio $x_1(t)/x_{1\,\max}$ and the relation dynamic tire load $x_2(t)/x_{2\,\max}$ are below 1, and the force of the actuator is below the maximum bound $u_{\max}$, which means the time-domain constraints are guaranteed by the designed controller.

From Fig. 3.3, we note that larger actuator forces are needed in the finite frequency control compared with that in the entire frequency control for the reason that the finite frequency control requires more force to match the finite frequency features. However, maybe it is worthwhile to conduct this in exchange for the advantages of

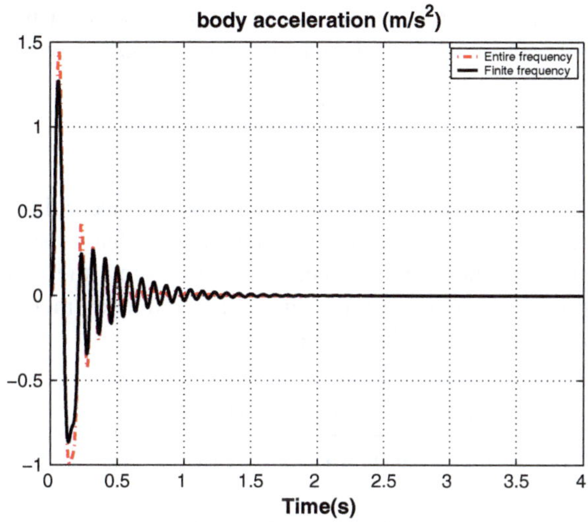

Fig. 3.2 Time-domain response of body vertical acceleration for active suspension system

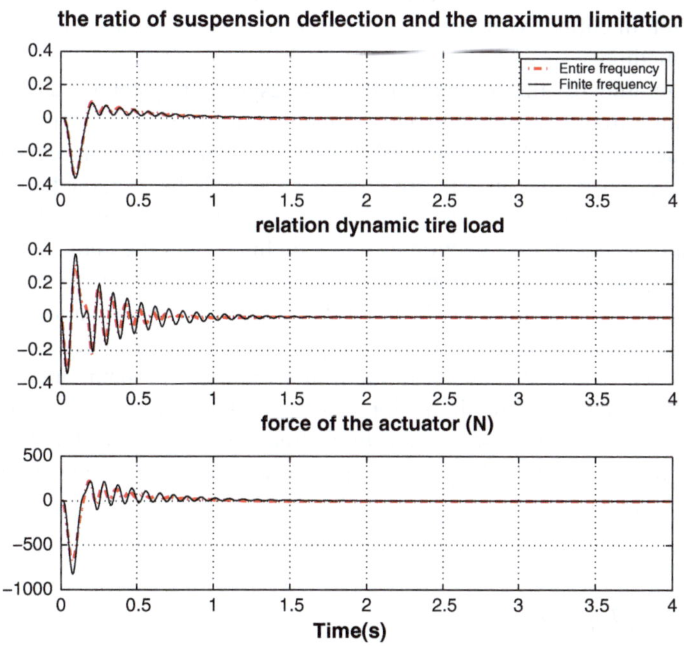

Fig. 3.3 Time-domain response of constraints for active suspension system

finite frequency control. Actuator power consumption is another important issue in automotive active control. We can calculate the actuator output average power by the formulation:

$$P = \frac{1}{T}\int_0^T (u(t)\cdot s(t))dt,$$

where "$s(t)$" respects the displacement of actuator, and according to the installed location, the displacement of actuator is equivalent to that of suspension, that is $s(t) = x_1(t)$, and T is the integral time. In order to show the comparison of the power consumptions between the finite and entire frequency methods, the ratio of the two kinds of powers is calculated, that is:

$$\delta = \frac{P_f}{P_e} = \frac{\frac{1}{T_f}\int_0^{T_f}(u_f(t)\cdot s_f(t))dt}{\frac{1}{T_e}\int_0^{T_e}(u_e(t)\cdot s_e(t))dt} = 0.7859,$$

where P_f and P_e respect the power consumptions of finite frequency control and entire frequency control, respectively. The above calculation implies the power consumption of actuator in finite frequency control is smaller than the counterpart (power consumption in entire frequency control), which further increases the feasibility of proposed method.

In the literature [8], the H_∞ control of active suspension systems is also considered over the finite frequency domain, and the method used to deal with the problem of finite frequency is to add some weighting functions to the active suspension systems and then the design parameters are chosen such that the weighted system norm is small. This weighting method is effective. However, this method is based on the appropriate weighting function as a precondition, and the choice of weighting function is quite time-consuming, especially when the designer has to shoot for a good trade-off between the complexity of the weights and the accuracy in capturing desired specifications. In this section, we provide a more reliable and convenient method to deal with the problem in the finite frequency domain, and avoid using the weighting function. Our simulation results validate that the ride comfort of the closed-loop system composed of the finite frequency controller has been greatly improved, and meanwhile the performance constraints are guaranteed within their allowable bounds.

3.1.3 Conclusion

This section has investigated the problem of H_∞ control with time domain constraints for active vehicle suspension systems in finite frequency domain. By the Generalized Kalman–Yakubovich–Popov lemma, the ride comfort has been improved by minimizing the H_∞ norm in specific frequency band, while the time-domain constraints have also been guaranteed in the framework of linear matrix inequality optimization. Analysis and simulation results for a quarter-car model has shown the effectiveness of the proposed approach.

3.2 Dynamic Output Feedback H_∞ Control for Active Suspensions

When all the states are on-line measurable, state feedback is an acceptable choice, as it can make use of full information, and thus the closed-loop performance can be enhanced to its full potential. However, state feedback control depends on the premise that all the state variables are on-line measurable, which leads into higher cost and additional complexity. In terms that not all the state variables can be measured on-line, output feedback control effects according to part of the measured states [9]. In other words, output feedback strategy requires less sensors, compared with the state feedback counterparts.

This section investigates the problem of H_∞ control for active suspension systems via dynamic output feedback control. By using the generalized Kalman–Yakubovich–Popov (KYP) lemma, the H_∞ norm from the disturbance to the controlled output is decreased over the chosen frequency band between which the human body is extremely sensitive to the vibration, to improve the ride comfort. In addition, the suspension deflection is limited within its allowed range to match the mechanical structure of the suspension. Considering the quarter-car suspension model, a dynamic output feedback controller is designed, where an effective multiplier expansion is used to convert the controller design to a convex optimization problem. Finally, a practical example is employed to illustrate the effectiveness of the proposed method.

Problem Formulation

By considering the vertical dynamics and taking into account the vehicle's symmetry, a suspension can be reduced to a quarter-car model illustrated in Chap. 1, as described in Fig. 1.2. Therefore, define state variables $\zeta(t)$, and output $z_{o1}(t)$, $z_{o2}(t)$

$$\begin{aligned} &\zeta_1(t) = z_s(t) - z_u(t),\ \zeta_1(t) = z_s(t) - z_u(t),\\ &\zeta_3(t) = \dot{z}_s(t),\ \zeta_4(t) = z_s(t) - \dot{z}_u(t), \end{aligned} \tag{3.33}$$

$$z_{o1}(t) = \ddot{z}_s(t),\ z_{o2}(t) = [\ \tfrac{z_s(t)-z_u(t)}{x_{1\max}}\ \ \tfrac{z_u(t)-z_r(t)}{x_2\max}\ ]^T, \tag{3.34}$$

and the vehicle suspension control system can be described as:

$$\begin{aligned} \dot{\zeta}(t) &= A\zeta(t) + Bu(t) + B_w w(t),\\ z_{o1}(t) &= C_1\zeta(t) + D_1 u(t),\\ z_{o2}(t) &= C_2\zeta(t),\\ y(t) &= C\zeta(t), \end{aligned} \tag{3.35}$$

where matrixes A, B, B_w, C_1, C_2, D_1 are defined in (2.9), and $y(t)$ is the measurable outputs. By choosing the proper matrix C, we can determine the state variables which the sensors can measure.

Roughly speaking, though state feedback control is a powerful strategy, it is based on the premise that all the state variables are online measurable, which sometimes introduces higher cost and additional complexity by measuring all the states. In the cases where not all the state variables can be measured on-line, output feedback control is an alternative, which can conduct effective control according to part of the measured states. In other words, output feedback strategy requires less sensors, compared with the state feedback counterparts, and has been investigated in many studies. In addition, the human body has different responses to different frequency vibrations, where vibrations over frequency 4–8 Hz are the major sources of the discomfort. A finite frequency H_∞ controller will be designed via dynamic output feedback control.

In this section, we design a dynamic output feedback H_∞ controller with the form:

$$\begin{aligned} \dot{\eta}(t) &= A_K\eta(t) + B_K y(t), \\ u(t) &= C_K\eta(t) + D_K y(t). \end{aligned} \tag{3.36}$$

where A_K, B_K, C_K, and D_K are dynamic feedback gain matrixes. According to the performance requirements, our goal can be summed up as follows:

$$\sup_{\varpi_1<\omega<\varpi_2} \|G(j\omega)\|_\infty < \gamma, \tag{3.37}$$

$$|\{z_{o2}(t)\}_i| \leq 1, \; i = 1, 2, \tag{3.38}$$

where $G(j\omega)$ represents the transfer function of the closed-loop systems from the disturbance input $w(t)$ to the controlled output $z_{o1}(t)$, and ϖ_1, ϖ_2 represent the lower and upper bounds of the chosen frequency.

3.2.1 Dynamic Output Feedback Controller Design

This subsection is devoted to the problem of finite frequency H_∞ controller design for the active suspension system. We are interested in designing a dynamic output feedback controller, such that the H_∞ norm of the closed-loop system is minimized over the chosen frequency range, while respecting the constraints in (3.38) within bound.

Substituting (3.36) into (3.35), and defining $x(t) = \left[\, \zeta^{\mathrm{T}}(t) \; \eta^{\mathrm{T}}(t) \,\right]^{\mathrm{T}}$, the closed-loop system admits the realization

$$\begin{aligned}\dot{x}(t) &= \bar{A}x(t) + \bar{B}w(t),\\ z_{o1}(t) &= \bar{C}_1x(t),\\ z_{o2}(t) &= \bar{C}_2x(t),\end{aligned} \tag{3.39}$$

where

$$\begin{aligned}\bar{A} &= \begin{bmatrix} A + BD_KC & BC_K \\ B_KC & A_K \end{bmatrix}, \ \bar{B} = \begin{bmatrix} B_w \\ 0 \end{bmatrix},\\ \bar{C}_1 &= \begin{bmatrix} C_1 & 0 \end{bmatrix}, \ \bar{C}_2 = \begin{bmatrix} C_2 & 0 \end{bmatrix},\end{aligned} \tag{3.40}$$

The transfer function of the closed-loop system from the disturbance input $w(t)$ to the controlled output $z_{o1}(t)$ is defined as follows:

$$G(j\omega) = \bar{C}_1(j\omega I - \bar{A})^{-1}\bar{B}.$$

Entire Frequency Case

In order to highlight the advantages of the finite frequency controller, we design another dynamic output feedback controller in the entire frequency domain, based on the method proposed in the literature [10]. In this subsection, a dynamic output feedback controller is designed over the entire frequency range, so that the closed-loop system in (3.39) is asymptotically stable, and satisfies

$$\sup_{-\infty<\omega<+\infty} \|G(j\omega)\|_\infty < \gamma, \tag{3.41}$$

while respecting the constraints in (3.38) within bound. The performance index (3.41) can be further expressed as: under zero initial condition, the closed-loop system guarantees that $\|z_1\|_2 < \gamma \|w\|_2$ for all nonzero $w \in L_2[0, \infty)$.

Theorem 3.7 *Let positive scalars ρ, γ be given. If there exists symmetric matrix $P_c > 0$ satisfying*

$$\begin{bmatrix} [P_c\bar{A}]_s & P_c\bar{B} & \bar{C}_1^T \\ * & -\gamma^2 I & 0 \\ * & * & -I \end{bmatrix} < 0, \tag{3.42}$$

$$\begin{bmatrix} -I & \sqrt{\rho}\{\bar{C}_2\}_i \\ * & -P_c \end{bmatrix} < 0, \tag{3.43}$$

then a stabilizing dynamic output feedback controller in the form of (3.36) exists, such that

(1) the closed-loop system in (3.39) is asymptotically stable;

(2) under zero initial condition, the closed-loop system guarantees that $\|z_{o1}\|_2 < \gamma \|w\|_2$ for all nonzero $w \in L_2[0, \infty)$;

(3) the constraint in (3.38) is guaranteed with the disturbance energy under the bound $w_{\max} = (\rho - V(0))/\gamma^2$.

Proof Since the results can easily be obtained, the proof has been omitted here. □

Hereafter, we will show how to transform (3.42) and (3.43) into the forms which can be solved directly. Partition the matrix P_c and its inverse P_c^{-1} in the form

$$P_c = \begin{bmatrix} Y_c & N_c \\ N_c^{\mathrm{T}} & \# \end{bmatrix}, \quad P_c^{-1} = \begin{bmatrix} X_c & M_c \\ M_c^{\mathrm{T}} & \# \end{bmatrix}, \tag{3.44}$$

where "#" represents this position can be arbitrary. From the literature [10], there is no loss of generality in assuming that N_c and M_c are invertible. Define

$$\Delta_{c1} = \begin{bmatrix} X_c & I \\ M_c^{\mathrm{T}} & 0 \end{bmatrix}, \quad \Delta_{c2} = \begin{bmatrix} I & Y_c \\ 0 & N_c^{\mathrm{T}} \end{bmatrix},$$

and note that

$$P_c \Delta_{c1} = \Delta_{c2}. \tag{3.45}$$

Define

$$J_{c1} = \mathrm{diag}\{\Delta_{c1},\ I,\ I\},\ J_{c2} = \mathrm{diag}\{I,\ \Delta_{c1}\}.$$

Pre- and post-multiplying (3.42) and (3.43) by J_{c1} and J_{c2} and their transposes, respectively, and defining

$$\begin{aligned}
\bar{A}_e &= \Delta_{c1}^{\mathrm{T}} P_c \bar{A} \Delta_2 = \begin{bmatrix} AX_c + B\hat{C}_c & A + B\hat{D}_c C \\ \hat{A}_c & Y_c A + \hat{B}_c C \end{bmatrix}, \\
\bar{B}_e &= \Delta_{c1}^{\mathrm{T}} P_c \bar{B} = \begin{bmatrix} B_1 \\ Y_c B_1 \end{bmatrix}, \\
\bar{C}_{e1} &= \bar{C}_1 \Delta_{c1} = \begin{bmatrix} C_1 X_c + D_1 \hat{C}_c & C_1 + D_1 \hat{D}_c C \end{bmatrix}, \\
\bar{C}_{e2} &= \bar{C}_2 \Delta_{c1} = \begin{bmatrix} C_2 X_c & C_2 \end{bmatrix}, \\
\bar{P}_c &= \Delta_{c1}^{\mathrm{T}} P_c \Delta_{c1} = \begin{bmatrix} X_c & I \\ I & Y_c \end{bmatrix},
\end{aligned}$$

with the following linearizing changes of variables:

$$\begin{aligned}
\hat{A}_c &= Y_c A X_c + Y_c B \hat{D}_c C X_c + N_c B_K C X_c + Y_c B C_K M_c^{\mathrm{T}} + N_c A_K M_c^{\mathrm{T}}, \\
\hat{B}_c &= Y_c B \hat{D}_c + N_c B_K, \\
\hat{C}_c &= C_K M_c^{\mathrm{T}} + \hat{D}_c C X_c, \\
\hat{D}_c &= D_K,
\end{aligned} \tag{3.46}$$

we can give the following corollary:

Theorem 3.8 *Let scalar* $\gamma > 0$ *be given. If there exist matrices* $Y_c > 0,\ X_c > 0$ *and general matrices* $\hat{A}_c,\ \hat{B}_c,\ \hat{C}_c,\ \hat{D}_c$ *satisfying*

$$\begin{bmatrix} [\bar{A}_e]_s & \bar{B}_e & \bar{C}_{e1}^T \\ * & -\gamma^2 I & 0 \\ * & * & -I \end{bmatrix} < 0, \tag{3.47}$$

$$\begin{bmatrix} -I & \sqrt{\rho}\bar{C}_{e2} \\ * & -\bar{P}_c \end{bmatrix} < 0. \tag{3.48}$$

then a stabilizing dynamic output feedback controller in the form of (3.36) exists, such that

(1) the closed-loop system in (3.39) is asymptotically stable;

(2) under zero initial condition, the closed-loop system guarantees that $\|z_{o1}\|_2 < \gamma\,\|w\|_2$ *for all nonzero* $w \in L_2\,[0,\infty)$*;*

(3) the constraint in (3.38) is guaranteed with the disturbance energy under the bound $w_{\max} = (\rho - V(0))/\gamma^2$.

Moreover, if inequalities (3.47) and (3.48) have a feasible solution, then we will compute the controller by

$$\begin{aligned} A_K &= N_c^{-1}[\hat{A}_c - Y_c A X_c - Y_c B \hat{D}_c C X_c - N_c B_K C X_c - Y_c B C_K M_c^T]M_c^{-T}, \\ B_K &= N_c^{-1}(\hat{B}_c - Y_c B \hat{D}_c), \\ C_K &= (\hat{C}_c - \hat{D}_c C X_c)M_c^{-T}, \\ D_K &= \hat{D}_c. \end{aligned} \tag{3.49}$$

Remark 3.9 The matrices N_c and M_c, which cannot be directly obtained by the Corollary 3.8, should satisfy

$$N_c M_c^{\mathrm{T}} = I - Y_c X_c.$$

Here, we obtain the two invertible matrices by using the singular value decomposition approach.

Finite Frequency Case

A dynamic output feedback controller is designed in the finite frequency band, so that the closed-loop system in (3.39) is asymptotically stable, and satisfies

$$\sup_{\varpi_1<\omega<\varpi_2} \|G(j\omega)\|_\infty < \gamma, \tag{3.50}$$

while respecting the constraint in (3.38) within bound.

Theorem 3.10 *Give positive scalars* γ, η, ρ *and let a dynamic output feedback controller in the form of (3.36) be given. The closed-loop system in (3.39) is asymptotically stable, and satisfies* $\sup_{\varpi_1<\omega<\varpi_2} \|G(j\omega)\|_\infty < \gamma$*, while respecting the constraint in (3.38) with the disturbance energy under the bound* $w_{\max} = (\rho - V(0))/\eta$*, if there exist symmetric matrices* $P,\ P_s > 0,\ Q > 0$ *and general matrix* W *satisfying*

$$\begin{bmatrix} [W]_s & -W^T\bar{A}+P_s & -W^T & -W^T\bar{B} \\ * & -P_s & 0 & 0 \\ * & * & -P_s & 0 \\ * & * & * & -\eta I \end{bmatrix} < 0, \tag{3.51}$$

$$\begin{bmatrix} -\varpi_1\varpi_2 Q - \left[\bar{A}^T W\right]_s & P - j\varpi_c Q + W^T & -W^T\bar{B} & \bar{C}_1^T \\ P + j\varpi_c Q + W & -Q & 0 & 0 \\ -\bar{B}^T W & 0 & -\gamma^2 I & 0 \\ \bar{C}_1 & 0 & 0 & -I \end{bmatrix} < 0, \tag{3.52}$$

$$\begin{bmatrix} -I & \sqrt{\rho}\{C_2\}_i \\ * & -P_s \end{bmatrix} < 0, \tag{3.53}$$

where $\varpi_c = (\varpi_1 + \varpi_2)/2$ *is a given scalar.*

Proof By using Schur complement, inequality (3.51) is equivalent to

$$\begin{bmatrix} \frac{1}{\eta}W^{\mathrm{T}}\bar{B}\bar{B}^{\mathrm{T}}W + W^{\mathrm{T}}P_s^{-1}W + [W]_s & -W^{\mathrm{T}}\bar{A} + P_s \\ * & -P_s \end{bmatrix} < 0. \tag{3.54}$$

Performing the congruence transformation to inequality (3.54) by $\mathrm{diag}\left\{-W^{-1}, P_s^{-1}\right\}$, with $W := -X^{-1}$, inequality (3.54) can be transformed to the following inequality:

$$\begin{bmatrix} \frac{1}{\eta}\bar{B}\bar{B}^{\mathrm{T}} + P_s^{-1} - [X]_s & \bar{A}P_s^{-1} + X^{\mathrm{T}} \\ * & -P_s^{-1} \end{bmatrix} < 0. \tag{3.55}$$

By using Lemma 3.3, inequality (3.55) is equivalent to

$$\bar{A}P_s^{-1} + P_s^{-1}\bar{A}^{\mathrm{T}} + \frac{1}{\eta}\bar{B}\bar{B}^{\mathrm{T}} < 0,$$

with $\Psi = \frac{1}{\eta}\bar{B}\bar{B}^{\mathrm{T}}$ and $S^{\mathrm{T}} = AP_s^{-1}$. Clearly, we have

$$\bar{A}^{\mathrm{T}}P_s + P_s\bar{A} + \frac{1}{\eta}P_s\bar{B}\bar{B}^{\mathrm{T}}P_s < 0, \tag{3.56}$$

which can guarantee $\bar{A}^{\mathrm{T}}P_s + P_s\bar{A} < 0$. From the standard Lyapunov theory for continuous-time linear system, the closed-loop system (3.39) is asymptotically stable with $w(t) = 0$.

Note that (3.52) is equivalent to

$$\begin{bmatrix} I & F_B \end{bmatrix} \Omega \begin{bmatrix} I & F_B \end{bmatrix}^{\mathrm{T}} + [F_A W R]_s < 0, \tag{3.57}$$

where

$$F_A = \begin{bmatrix} -\bar{A}^{\mathrm{T}} \\ I \\ -\bar{B}^{\mathrm{T}} \end{bmatrix}, \quad F_B = \begin{bmatrix} \bar{C}_1^{\mathrm{T}} \\ 0 \\ 0 \end{bmatrix}, \quad R = \begin{bmatrix} I \\ 0 \\ 0 \end{bmatrix}^{\mathrm{T}}, \quad \Omega = T \begin{bmatrix} \Phi \otimes P + \Psi \otimes Q & 0 \\ 0 & \Pi \end{bmatrix} T^{\mathrm{T}},$$

$$\Pi = \begin{bmatrix} I & 0 \\ 0 & -\gamma^2 I \end{bmatrix}, \quad \Phi = \begin{bmatrix} 0 & 1 \\ 1 & 0 \end{bmatrix}, \quad \Psi = \begin{bmatrix} -1 & j\varpi_c \\ -j\varpi_c & -\varpi_1\varpi_2 \end{bmatrix},$$

with the permutation matrix T such as $\begin{bmatrix} M_1 & M_2 & M_3 & M_4 \end{bmatrix} T = \begin{bmatrix} M_2 & M_1 & M_4 & M_3 \end{bmatrix}$. From Lemma 3.2, (3.57) is equivalent to

$$F_A^{\perp} \begin{bmatrix} I & F_B \end{bmatrix} \Omega \begin{bmatrix} I & F_B \end{bmatrix}^{\mathrm{T}} F_A^{\perp^{\mathrm{T}}} < 0, \tag{3.58}$$

$$R^{\mathrm{T}^{\perp}} \begin{bmatrix} I & F_B \end{bmatrix} \Omega \begin{bmatrix} I & F_B \end{bmatrix}^{\mathrm{T}} R^{\mathrm{T}^{\perp^{\mathrm{T}}}} < 0, \tag{3.59}$$

where inequality (3.58) holds if and only if

$$F^{\mathrm{T}} \Omega F < 0, \tag{3.60}$$

with

$$F = \begin{bmatrix} I & \bar{A}^{\mathrm{T}} & 0 & \bar{C}_1^{\mathrm{T}} \\ 0 & \bar{B}^{\mathrm{T}} & I & 0 \end{bmatrix}^{\mathrm{T}}.$$

The inequality in (3.60) is equivalent to

$$\begin{bmatrix} \bar{A} & \bar{B} \\ I & 0 \\ \bar{C}_1 & 0 \\ 0 & I \end{bmatrix}^{\mathrm{T}} \begin{bmatrix} \Phi \otimes P + \Psi \otimes Q & 0 \\ 0 & \Pi \end{bmatrix} \begin{bmatrix} \bar{A} & \bar{B} \\ I & 0 \\ \bar{C}_1 & 0 \\ 0 & I \end{bmatrix} < 0, \tag{3.61}$$

which is further equivalent to

$$\begin{bmatrix} \bar{A} & \bar{B} \\ I & 0 \end{bmatrix}^{\mathrm{T}} (\Phi \otimes P + \Psi \otimes Q) \begin{bmatrix} \bar{A} & \bar{B} \\ I & 0 \end{bmatrix} + \begin{bmatrix} \bar{C}_1 & 0 \\ 0 & I \end{bmatrix}^{\mathrm{T}} \Pi \begin{bmatrix} \bar{C}_1 & 0 \\ 0 & I \end{bmatrix} < 0. \tag{3.62}$$

By using Lemma 3.1, the above inequality is equivalent to

$$\begin{bmatrix} (j\omega I - \bar{A})^{-1} \bar{B} \\ I \end{bmatrix}^{*} \begin{bmatrix} \bar{C}_1 & 0 \\ 0 & I \end{bmatrix}^{*} \Pi \begin{bmatrix} \bar{C}_1 & 0 \\ 0 & I \end{bmatrix} \begin{bmatrix} (j\omega I - \bar{A})^{-1} \bar{B} \\ I \end{bmatrix} < 0, \; \varpi_1 < \omega < \varpi_2, \tag{3.63}$$

which is equivalent to the finite frequency H_∞ performance index inequality

$$\sup_{\varpi_1 < \omega < \varpi_2} \| G(j\omega) \|_\infty < \gamma. \tag{3.64}$$

Denote $V(t) = x^{\mathrm{T}}(t)P_s x(t)$ as the energy function, whose derivative is obtained as

$$\dot{V}(t) = 2x^{\mathrm{T}}(t)P_s\bar{A}x(t) + 2x^{\mathrm{T}}(t)P_s\bar{B}w(t).$$

Noting that

$$2x^{\mathrm{T}}(t)P_s\bar{B}w(t) \leq \frac{1}{\eta}x^{\mathrm{T}}(t)P_s\bar{B}\bar{B}^{\mathrm{T}}P_s x(t) + \eta w^{\mathrm{T}}(t)w(t), \forall\eta > 0,$$

we have

$$\dot{V}(t) \leq x^{\mathrm{T}}(t)(\bar{A}^{\mathrm{T}}P_s + P_s\bar{A} + \frac{1}{\eta}P_s\bar{B}\bar{B}^{\mathrm{T}}P_s)x(t) + \eta w^{\mathrm{T}}(t)w(t).$$

According to inequality (3.56), we have

$$\dot{V}(t) \leq \eta w^{\mathrm{T}}(t)w(t).$$

Integrating both sides of the above inequality from 0 to t results in

$$V(t) - V(0) \leq \eta\int_0^t w^{\mathrm{T}}(t)w(t)dt \leq \eta\,\|w\|_2^2 = \eta w_{\max}.$$

This shows that

$$x^{\mathrm{T}}(t)P_s x(t) \leq V(0) + \eta w_{\max} = \rho. \tag{3.65}$$

Consider

$$\max_{t\geq 0}|\{z_{o2}(t)\}_i|^2 = \max_{t\geq 0}\left\|x^T(t)\{C_2\}_i^T\{C_2\}_i x(t)\right\|_2, \; i = 1, 2.$$

Using the transformation $\bar{x}(t) = \bar{P}_s^{\frac{1}{2}}x(t)$, from inequality (3.65) it follows that

$$\bar{x}^{\mathrm{T}}(t)\bar{x}(t) \leq \rho.$$

Hence,

$$\begin{aligned}\max_{t\geq 0}|\{z_{o2}(t)\}_i|^2 &= \max_{t\geq 0}\left\|x^T(t)\{C_2\}_i^T\{C_2\}_i x(t)\right\|_2 \\ &\leq \rho\cdot\lambda_{\max}(P_s^{-\frac{1}{2}}\{C_2\}_i^T\{C_2\}_i P_s^{-\frac{1}{2}}), \; i = 1, 2,\end{aligned} \tag{3.66}$$

where $\lambda_{\max}(\cdot)$ represents the maximum eigenvalue. Then, the constraint in (3.38) holds if

$$\rho P_s^{-\frac{1}{2}}\{C_2\}_i^T\{C_2\}_i P_s^{-\frac{1}{2}} < I, \; i = 1, 2, \tag{3.67}$$

which, by Schur complement, is equivalent to (3.53). The proof is completed. □

Expressions in Theorem 3.10 are non-convex due to the product terms of the multiplier W, the controller parameters and coefficient matrices. In order to solve this problem, we carry on the following transformation.

In accordance with the partition of $\bar{A}$ in (3.40), we introduce a partition of W and its inverse W^{-1} in the form:

$$W = \begin{bmatrix} X & Y \\ U & V \end{bmatrix}, \; W^{-1} = \begin{bmatrix} M & G \\ H & L \end{bmatrix}. \tag{3.68}$$

From the literature [6], there is no loss of generality in assuming that U and H are invertible. Define

$$\Delta_1 = \begin{bmatrix} X & I \\ U & 0 \end{bmatrix}, \; \Delta_2 = \begin{bmatrix} I & M \\ 0 & H \end{bmatrix},$$

and note that

$$W\Delta_2 = \Delta_1. \tag{3.69}$$

Define

$$J_1 = \text{diag}\left\{\Delta_2^{\mathrm{T}},\ \Delta_2^{\mathrm{T}},\ \Delta_2^{\mathrm{T}},\ I\right\},\ J_2 = \text{diag}\left\{\Delta_2^{\mathrm{T}},\ \Delta_2^{\mathrm{T}},\ I,\ I\right\},\ J_3 = \text{diag}\left\{I,\ \Delta_2^{\mathrm{T}}\right\}.$$

Pre- and post-multiplying (3.51), (3.52) and (3.53) by J_1, J_2 and J_3 and their transposes, respectively, and defining

$$\begin{aligned}
\bar{Q} &= \Delta_2^{\mathrm{T}} Q \Delta_2,\ \bar{P} = \Delta_2^{\mathrm{T}} P \Delta_2,\ \bar{P}_s = \Delta_2^{\mathrm{T}} P_s \Delta_2,\\
\hat{A} &= \Delta_2^{\mathrm{T}} W^{\mathrm{T}} \bar{A} \Delta_2 = \begin{bmatrix} X^{\mathrm{T}} A + \hat{B}_K C & \hat{A}_K \\ A + B\hat{D}_K C & AM + B\hat{C}_K \end{bmatrix},\\
\hat{B} &= \Delta_2^{\mathrm{T}} W^{\mathrm{T}} \bar{B} = \begin{bmatrix} X^{\mathrm{T}} B_w \\ B_w \end{bmatrix},\\
\hat{C}_1 &= \bar{C}_1 \Delta_2 = \begin{bmatrix} C_1 + D_1 \hat{D}_K C & C_1 M + D_1 \hat{C}_K \end{bmatrix},\\
\hat{C}_2 &= \bar{C}_2 \Delta_2 = \begin{bmatrix} C_2 & C_2 M \end{bmatrix},\\
\bar{W} &= \Delta_2^{\mathrm{T}} W \Delta_2 = \begin{bmatrix} X^{\mathrm{T}} & Z \\ I & M \end{bmatrix},
\end{aligned}$$

with the following linearizing changes of variables:

$$\begin{aligned}
\hat{A}_K &= X^{\mathrm{T}} AM + X^{\mathrm{T}} B D_K C M + U^{\mathrm{T}} B_K C M + X^{\mathrm{T}} B C_K H \\
&\quad + U^{\mathrm{T}} A_K H,\\
\hat{B}_K &= X^{\mathrm{T}} B D_K + U^{\mathrm{T}} B_K,\\
\hat{C}_K &= C_K H + D_K C M,\\
\hat{D}_K &= D_K,\\
Z &= X^{\mathrm{T}} M + U^{\mathrm{T}} H,
\end{aligned} \tag{3.70}$$

we can give the following theorem.

Theorem 3.11 *Give positive scalars γ, η, ρ. A dynamic output feedback controller in the form of (3.36) exists, such that the closed-loop system in (3.39) is asymptotically stable, and satisfies $\sup_{\varpi_1<\omega<\varpi_2} \|G(j\omega)\|_\infty < \gamma$, while respecting the constraint in (3.38) with the disturbance energy under the bound $w_{\max} = (\rho - V(0))/\eta$, if there exist symmetric matrices $\bar{P}$, $\bar{P}_s > 0$, $\bar{Q} > 0$ and general matrices $\bar{W}$, $\hat{A}_k$, $\hat{B}_k$, $\hat{C}_k$, $\hat{D}_k$, M, X and Z satisfying*

$$\begin{bmatrix} [\bar{W}]_s & -\hat{A}+\bar{P}_s & -\bar{W}^T & -\hat{B} \\ * & -\bar{P}_s & 0 & 0 \\ * & * & -\bar{P}_s & 0 \\ * & * & * & -\eta I \end{bmatrix} < 0, \tag{3.71}$$

$$\begin{bmatrix} -\varpi_1\varpi_2\bar{Q} - [\hat{A}]_s & \bar{P} - j\varpi_c\bar{Q} + \bar{W}^T & -\hat{B} & \hat{C}_1^T \\ \bar{P} + j\varpi_c\bar{Q} + \bar{W} & -\bar{Q} & 0 & 0 \\ -\hat{B}^T & 0 & -\gamma^2 I & 0 \\ \hat{C}_1 & 0 & 0 & -I \end{bmatrix} < 0, \tag{3.72}$$

$$\begin{bmatrix} -I & \sqrt{\rho}\hat{C}_2 \\ * & -\bar{P}_s \end{bmatrix} < 0. \tag{3.73}$$

Moreover, if the above inequalities has a feasible solution, then the matrices $\hat{A}_k$, $\hat{B}_k$, $\hat{C}_k$, $\hat{D}_k$, M, X and Z can be obtained. According to (3.70)–(3.71), we will compute the controller by

$$\begin{aligned} A_K &= U^{-T}[\hat{A}_K - X^T A M - X^T B D_K C M - U^T B_K C M - X^T B C_K H]H^{-1}, \\ B_K &= U^{-T}(\hat{B}_K - X^T B D_K), \\ C_K &= (\hat{C}_K - D_K C M)H^{-1}, \\ D_K &= \hat{D}_K. \end{aligned} \tag{3.74}$$

Remark 3.12 Note that the linear matrix inequality in (3.72) has complex variables. The LMI in complex variables can be converted to an LMI of larger dimension in real variables. This means that inequality $S_1 + jS_2 < 0$ is equivalent to $\begin{bmatrix} S_1 & S_2 \\ -S_2 & S_1 \end{bmatrix} < 0$. Based on the above method, inequality (3.72) can be split into

$$\begin{bmatrix} -\varpi_1\varpi_2\bar{Q} - [\hat{A}]_s & \bar{P} + \bar{W}^T & -\hat{B} & \hat{C}_1^T \\ * & -\bar{Q} & 0 & 0 \\ * & * & -\gamma^2 I & 0 \\ * & * & * & -I \end{bmatrix} + j \begin{bmatrix} 0 & -\varpi_c\bar{Q} & 0 & 0 \\ \varpi_c\bar{Q} & 0 & 0 & 0 \\ 0 & 0 & 0 & 0 \\ 0 & 0 & 0 & 0 \end{bmatrix} < 0, \tag{3.75}$$

which can be solved by

$$\begin{bmatrix} -\varpi_1\varpi_2\bar{Q}-\left[\hat{A}\right]_s & \bar{P}+\bar{W}^{\mathrm{T}} & -\hat{B} & \hat{C}_1^{\mathrm{T}} & 0 & -\varpi_c\bar{Q} & 0 & 0 \\ * & -\bar{Q} & 0 & 0 & \varpi_c\bar{Q} & 0 & 0 & 0 \\ * & * & -\gamma^2 I & 0 & 0 & 0 & 0 & 0 \\ * & * & * & -I & 0 & 0 & 0 & 0 \\ * & * & * & * & -\varpi_1\varpi_2\bar{Q}-\left[\hat{A}\right]_s & \bar{P}+\bar{W}^{\mathrm{T}} & -\hat{B} & \hat{C}_1^{\mathrm{T}} \\ * & * & * & * & * & -\bar{Q} & 0 & 0 \\ * & * & * & * & * & * & -\gamma^2 I & 0 \\ * & * & * & * & * & * & * & -I \end{bmatrix} < 0. \tag{3.76}$$

Remark 3.13 When we calculate the controller, the matrices U and H, which cannot be directly obtained by the Theorem 3.11, are needed and they should be chosen such that

$$U^{\mathrm{T}}H = Z - X^{\mathrm{T}}M.$$

It is worth mentioning that the factorization of $U^{\mathrm{T}}H$ can always be achieved so that the invertible matrices U and H are deduced. In this section, the two invertible matrices are obtained by using the singular value decomposition approach.

3.2.2 *Simulation Verification*

In this subsection, we will apply the above approach to design a dynamic output feedback H_∞ controller in the finite frequency domain based on the active suspension model described in Sect. 3.2.1. The measure matrix is chosen as $C = \left[\, I_{3\times3}\ 0_{3\times1} \,\right]$. The parameters of the suspension system are the same as those applied in Sect. 2.1, which are listed in Table 2.1. The Maximums of the time-domain constraints and scalar values are listed in Table 3.2.

Firstly, the closed-loop system with a dynamic output feedback H_∞ controller in the finite frequency domain can be obtained, based on the proposed method in

Table 3.2 Maximums of the time-domain constraints and scalar values

The maximum suspension deflection $x_{1\max}$	0.1 m
The maximum tire deflection $x_{2\max}$	0.01764 m
The maximum active force $u_{\max}$	2500 N
The positive scalar ρ	1
The positive scalar η	10000
The lower bound ϖ_1	4 Hz
The upper bound ϖ_2	8 Hz

Sect. 3.2.1, and we denote this closed-loop system as system Σ_1 for brevity. After solving the matrix inequalities (3.71)–(3.73) for symmetric matrices $\bar{P}$, $\bar{P}_s > 0$, $\bar{Q} > 0$ and general matrices $\bar{W}$, $\hat{A}_k$, $\hat{B}_k$, $\hat{C}_k$, $\hat{D}_k$, M, X and Z with given scalars $\gamma > 0$ and $\varpi_1 = 8\pi$rad/s (4 Hz), $\varpi_2 = 16\pi$rad/s (8 Hz), $\eta = 10000$, the optimal guaranteed closed-loop H_∞ performance obtained is

$$\gamma_{\min} = 3.1718.$$

Then, the parameter matrices of the dynamic output feedback controller are obtained.

For subsequent comparison, we can get another closed-loop system with a dynamic output feedback H_∞ controller over the entire frequency range, according to the Theorem 3.8, and set it as system Σ_2. After solving the matrix inequalities in Theorem 3.8, we obtain the optimal guaranteed closed-loop H_∞ performance:

$$\gamma_{\min} = 4.8113.$$

The parameter matrices of the dynamic output feedback controller in entire frequency domain can also be solved.

In order to further illustrate the effectiveness of disturbance suppression over the frequency band 4–8 Hz, the curves of maximum singular values are drawn in Fig. 3.4, where the open-loop system (passive mode), the closed-loop system Σ_1 (active finite frequency mode) and the closed-loop system Σ_2 (active entire frequency mode), are compared. In Fig. 3.4, the dash/dash-dot/solid line represents the curve of maximum singular values in the open-loop system/system Σ_2/system Σ_1, respectively. From Fig. 3.4, we can see that the closed-loop system with finite frequency controller has the least value of H_∞ norm over the frequency range 4–8 Hz, compared with the passive system and the closed-loop system with an entire frequency controller,

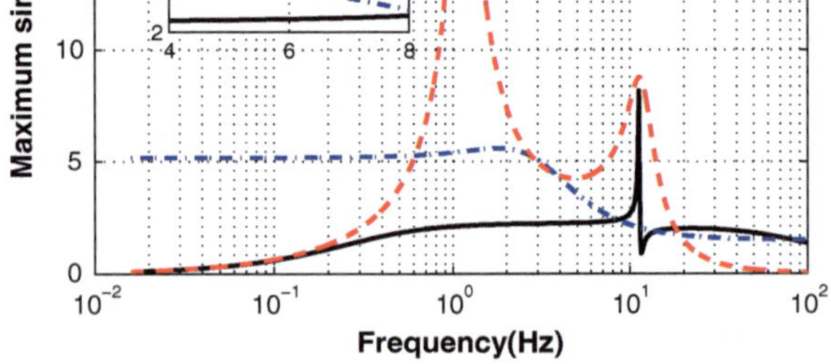

Fig. 3.4 The curves of maximum singular values (blue line: open-loop system; red line: system Σ_2; black line: system Σ_1)

which means an improved ride comfort has been achieved by the finite frequency controller.

Evaluation of the vehicle suspension performance is based on the examination of three response quantities, that is, the body acceleration of the specific frequency domain, the suspension stroke, and tire deflection. In order to evaluate the suspension characteristics with respect to the three performance requirements, both certain and random inputs are employed in this simulation.

Bump Road Inputs

It is assumed that the certain disturbance input has the same following form described as (3.32). The time-domain responses of body vertical acceleration for the active suspension systems are shown in Fig. 3.5, where the solid/dash-dot lines are the responses of body vertical acceleration with the finite/entire frequency controller, and the dash line respects the responses of the passive system. It is seen from this figure that the magnitudes for the body accelerations are significantly decreased for active suspensions. Also, the acceleration for the finite frequency controlled active suspension vanish faster than the other two suspensions (passive suspension and entire frequency controlled suspension). These results confirm the efficiency of the finite frequency controller. In particular, reduced acceleration indicates that the ride comfort is improved. In addition, Figs. 3.6 and 3.7 show that the ratio $z_1/x_{1\,\max}$ and $z_2/x_{2\,\max}$ is below 1, respectively, which means the time-domain constraints (the suspension stroke and tire deflection) are guaranteed by the designed controller in terms of the road input (3.32).

Random Road Inputs

Simulations are conducted under random road excitation in the form of white noise. The power spectra density of uneven road velocity input is given by (1.19) in Chap. 1:

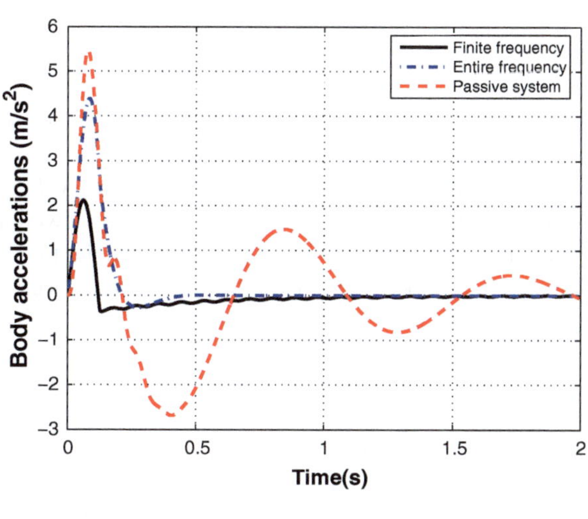

Fig. 3.5 The time-domain response of body acceleration

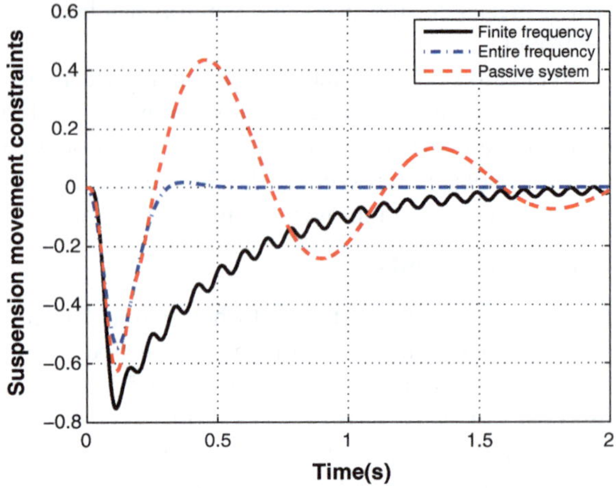

Fig. 3.6 Suspension movement constrains (the ratio $z_1/x_{1\max}$)

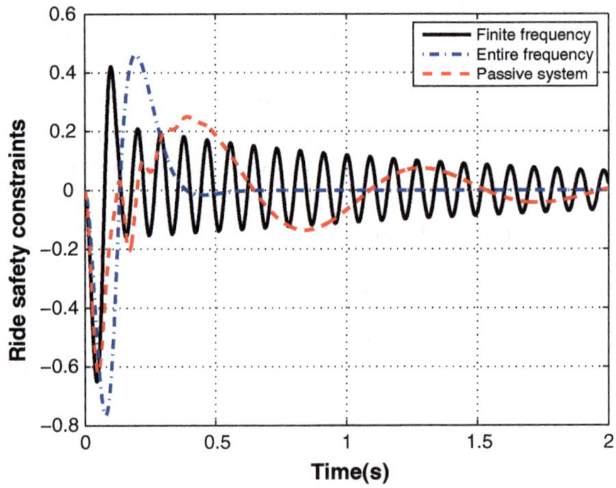

Fig. 3.7 Tire deflection constrains (the ratio $z_2/x_{2\max}$)

$$G_{\dot{q}}(f) = 4\pi^2 G_q(n_0) v n_0^2$$

which is only related to the vehicle forward velocity. When the vehicle forward velocity is fixed, the ground velocity can be viewed as a white-noise signal. Let the road class be C, with road roughness coefficient of $256 \times 10^{-6}\text{m}^3$, and the vehicle drives at the speed of 45 km/h. The power spectra density of the vehicle heave acceleration can be described as:

$$\phi_{z_1}(f) = \left|G_{w\to z_1}(j\omega)\right|^2 G_{\dot{q}}(f) \tag{3.77}$$

Select the road roughness as $G_q(n_0) = 16 \times 10^{-6}\text{m}^3$, $G_q(n_0) = 64 \times 10^{-6}\text{m}^3$, $G_q(n_0) = 256 \times 10^{-6}\text{m}^3$, and $G_q(n_0) = 1024 \times 10^{-6}\text{m}^3$, which are corresponded to A, B, C, and D Grade respectively according to ISO2361 standards, to generate

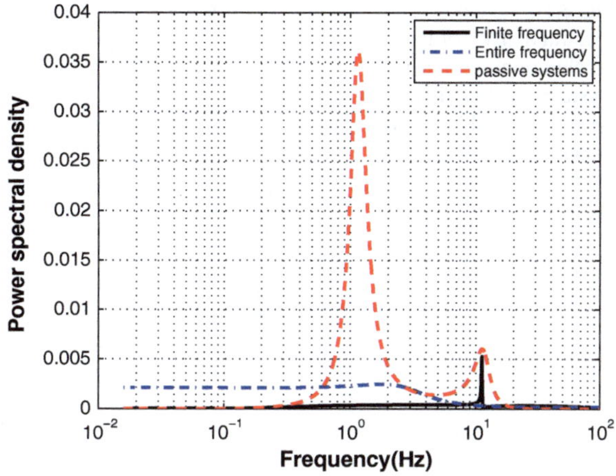

Fig. 3.8 The power spectral density of body acceleration in grade A (system Σ_1: blue line; system Σ_2: red line; passive system: red line)

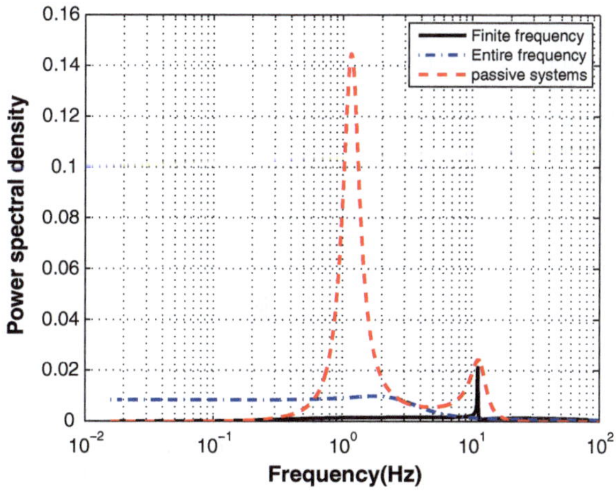

Fig. 3.9 The power spectral density of body acceleration in grade B (system Σ_1: blue line; system Σ_2: red line; passive system: red line)

the random road profile. Set the vehicle forward velocity as $v = 45$ km/h, and as expected, it is observed from Figs. 3.8, 3.9, 3.10 and 3.11 that the closed-loop system Σ_1 with finite frequency controller realizes a better ride comfort, compared with system Σ_2 and passive system over the frequency range 4–8 Hz (since the system Σ_1 has lower PSD body acceleration than system Σ_2, and smaller PSD body acceleration value results in better ride comfort), where PSD body acceleration can be calculated by

$$G_{z_1}(f) = |G(j\omega)|^2 G_{\dot{q}}(f). \tag{3.78}$$

With the series of simulations above, a fact is proved once again: in the selected frequency domain, the finite frequency method beats the entire frequency method in the capability of disturbance suppression. The reason for this is that the finite frequency approach concentrates control powers on the chosen frequency domain

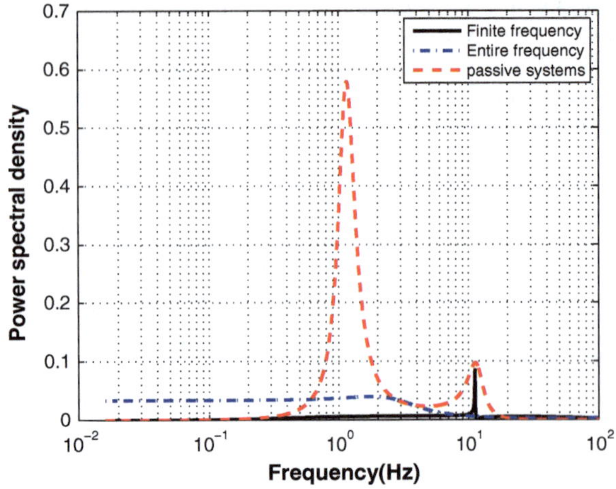

Fig. 3.10 The power spectral density of body acceleration in grade C (system Σ_1: blue line; system Σ_2: red line; passive system: red line)

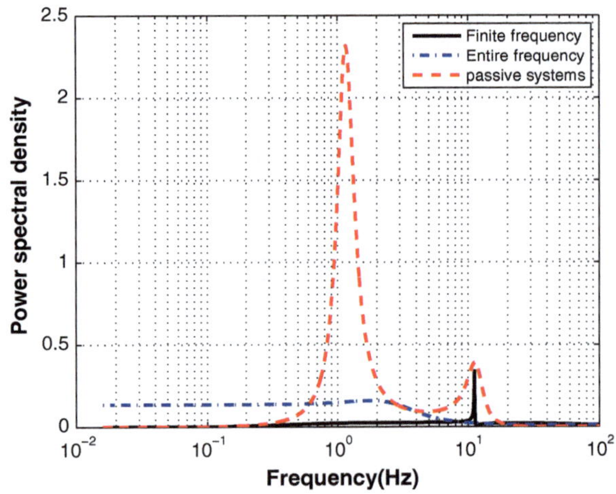

Fig. 3.11 The power spectral density of body acceleration in grade D (system Σ_1: blue line; system Σ_2: red line; passive system: red line)

and relaxes the restrictions of the other frequencies, by imposing the frequency band constraints in the performance indicators.

3.2.3 Conclusion

In this section, a dynamic output feedback H_∞ controller for active suspension system has been designed, which can improve ride comfort as much as possible. The key idea of designing the proposed controller is to use the generalized Kalman–Yakubovich–Popov (GKYP) lemma and the linearizing change of variables. In addition, the limited suspension stroke and dynamic tire deflection are guaranteed by considering these constraints in the controller design. The simulation results show that the finite frequency output feedback H_∞ controller achieves better disturbance attenuation for the concerned frequency range, and the performance constraints are also guaranteed.

References

1. M. Yamashita, K. Fujimori, K. Hayakawa, H. Kimura, Application of H_∞ control to active suspension systems. Automatica **30**(11), 1717–1729 (1994)
2. B. Hencey, A.G. Alleyne, A KYP lemma for LMI regions. IEEE Trans. Autom. Control. **52**(10), 1926–1930 (2007)
3. H. Khatibi, A. Karimi, R. Longchamp, Fixed-order controller design for polytopic systems using LMIs. IEEE Trans. Autom. Control. **53**(1), 428–434 (2008)
4. J. Collado, R. Lozano, R. Johansson, On Kalman–Yakubovich–Popov lemma for stabilizable systems. IEEE Trans. Autom. Control. **46**(7), 1089–1093 (2001)
5. T. Iwasaki, S. Hara, Generalized KYP lemma: unified frequency domain inequalities with design applications. IEEE Trans. Autom. Control. **50**(1), 41–59 (2005)
6. P. Apkarian, H.D. Tuan, J. Bernussou, Continuous-time analysis, eigenstructure assignment, and H_2 synthesis with enhanced Linear Matrix Inequalities (LMI) characterizations. IEEE Trans. Autom. Control. **42**(12), 1941–1946 (2001)
7. P. Gahinet, A. Nemirovskii, A.J. Laub, M. Chilali, *LMI Control Toolbox User's Guide* (The Math Works Inc., Natick, MA, 1995)
8. M. Yamashita, K. Fujimori, K. Hayakawa, H. Kimura, Application of H_∞ control to active suspension systems. Automatica **30**(11), 1717–1729 (1994)
9. H. Karimi, H. Gao, Mixed H_2/H_∞ output-feedback control of second-order neutral systems with time-varying state and input delays. ISA Trans. **47**(3), 311–324 (2008)
10. C. Scherer, P. Gahinet, M. Chilali, Multi-objective output-feedback control via LMI optimization. IEEE Trans. Autom. Control. **42**(7), 896–911 (1997)

Chapter 4
Constrained Active Suspension Control via Nonlinear Feedback Technology

In the above chapters, the linear quarter-car active suspension systems are studied based on H_∞ control approach over both entire frequency domain and finite frequency domain. For half-car model and full-car model, however, the vehicle suspensions are no longer easily simplified as linear systems, but nonlinear ones, because of the nonlinear characteristics of suspension components and the coupling between subsystems. As a consequence, some nonlinear control methods should be put forward to tackle the problem. In Sect. 4.1, terminal slide mode control is applied to realize finite time-stabilization which ensures that the tracking errors reach zero in finite time. Meanwhile, the singularity and chattering problems are resolved, and robustness and disturbance rejection properties of the system are good. In Sect. 4.2, when designing the adaptive backstepping controller, parameter uncertainties including spring nonlinearity and the piece-wise linear behavior of the damper are taken into consideration to form the basis of accurate control.

4.1 Finite-Time Stabilization for Vehicle Active Suspension Systems with Hard Constraints

This section presents the problem of finite-time stabilization for vehicle suspension systems with hard constraints based on terminal sliding-mode (TSM) control. As we know, one of the strong points of TSM control is its finite-time convergence to a given equilibrium of the system under consideration, which may be useful in specific applications. However, two main problems hindering the application of the TSM control are the singularity and chattering in TSM control systems. This section proposes a novel second-order sliding-mode algorithm to soften the switching control law. The effect of the equivalent low-pass filter can be properly controlled in the algorithm based on requirements. Meantime, since the derivatives of term with fractional power do not appear in the control law, the control singularity is avoided. Thus, a chattering-free TSM control scheme for suspension systems is proposed,

W. Sun et al., *Advanced Control for Vehicle Active Suspension Systems*, Studies in Systems, Decision and Control 204,
https://doi.org/10.1007/978-3-030-15785-2_4

which allows both the chattering and singularity problems to be resolved. Finally, the effectiveness of the proposed approach is illustrated by both theoretical analysis and comparative experiment results.

4.1.1 System Description and Preliminaries

System Description

In this section, the quarter-car model with nonlinearities shown in Fig. 4.1 is considered. Also, the effect of actuator dynamics is neglected and the actuator is modelled as an ideal force generator. The m_s and m_u represent the sprung mass and the unsprung mass of the suspension. The z_s and z_u are the vertical displacements of the sprung mass and the unsprung mass, respectively, and z_r is the vertical road profile. The tire is assured contract with the surface of the road when the vehicle is traveling. It is modeled as a linear spring producing the elasticity force F_t and a linear damper producing the damping force F_b. The passive components of the suspension system consist of a non-linear stiffening spring producing the force F_s and the piece-wise linear damper producing the force F_d; u denotes the active input of the suspension system.

The motion equation of a quarter-car suspension system may be established as following:

$$
\begin{aligned}
m_s\ddot{z}_s &= -F_d(\dot{z}_s, \dot{z}_u, t) - F_s(z_s, z_u, t) + u(t) + f_\Delta(t), \\
m_u\ddot{z}_u &= F_d(\dot{z}_s, \dot{z}_u, t) + F_s(z_s, z_u, t) - F_t(z_u, z_r, t) \\
&\quad - F_b(\dot{z}_u, \dot{z}_r, t) - u(t),
\end{aligned} \tag{4.1}
$$

where the variable f_Δ is taken into account for the modeling, which represents the system parameter uncertainties and the external disturbances of suspension components. It is to be noted that with a change in the number of passengers or the payload, the vehicle load will easily vary and this will accordingly change the vehicle mass m_s. So m_{s_0} is defined as a nominal parameter which is an crude estimation of the true parameters in (4.1), i.e., $m_s = m_{s_0} + \Delta m_{s_0}$. The main nonlinearity of the suspension system is the nonlinear force-velocity characteristic of the damper and spring. The characteristic of the primary damper has been identified to be piece-wise linear in the operating range. The tire force-deflection characteristic has been identified to be progressive but can be linearized in the operating point given by the static wheel load. The identified tire damping is comparably small, which is coherent with the literature (see [1]). The forces produced by the non-linear stiffening spring, the piece-wise linear damper and the tire can be calculated as:

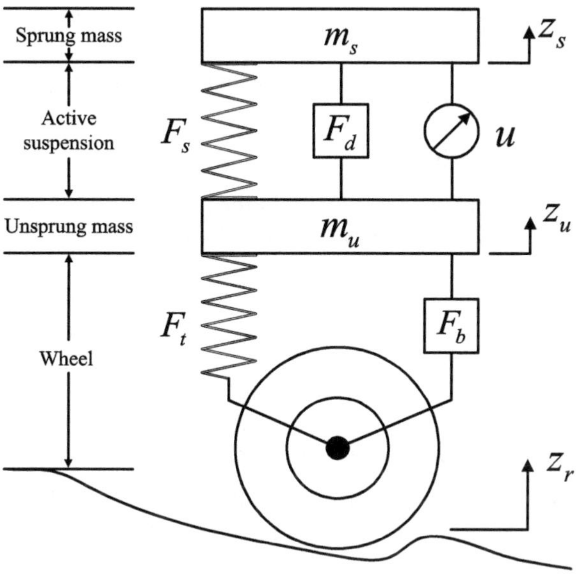

Fig. 4.1 The structure of quarter-car active suspension

$$F_s = k_{s_1}(z_s - z_u) + k_{n_1}(z_s - z_u)^3, \tag{4.2}$$

$$F_d = \begin{cases} b_e(\dot{z}_s - \dot{z}_u), \dot{z}_s - \dot{z}_u > 0, \\ b_c(\dot{z}_s - \dot{z}_u), \dot{z}_s - \dot{z}_u \leq 0, \end{cases} \tag{4.3}$$

$$F_t = k_f(z_u - z_r), \;\; F_b = b_f(\dot{z}_u - \dot{z}_r), \tag{4.4}$$

where k_{s_1} and k_{n_1} are the stiffness coefficient of the linear terms and the cubic terms, respectively; b_e and b_c are the damping coefficient for the extension movement and the compression movement, respectively; k_f, b_f are the stiffness and damping coefficients of the tires.

Let $x_1(t) = z_s(t)$ which denotes the displacement of the sprung, $x_2(t) = \dot{z}_s(t)$ the sprung mass speed, $x_3(t) = z_u(t)$ the displacement of the unsprung, and $x_4(t) = \dot{z}_u(t)$ the unsprung mass speed. The differential equations in (4.1) can be written in the following state space form:

$$\begin{aligned} \dot{x}_1 &= x_2, \\ m_{s_0}\dot{x}_2 &= \phi(x, t) + u + F_\Delta, \\ \dot{x}_3 &= x_4, \\ m_u\dot{x}_4 &= -\phi(x, t) - F_t - F_b - u, \end{aligned} \tag{4.5}$$

where $\phi(x, t) = -F_d - F_s$, u is a control law, $F_\Delta = f_\Delta - \Delta m_{s_0}\dot{x}_2$ is the lump disturbance and assumed to satisfy the following condition $|F_\Delta| \leq m_{s_0} l_d$, where $l_d > 0$ is a bounded constant. In addition, we assume that the derivative of F_Δ in system (4.5) is bounded:

$$|\dot{F}_{\Delta}| \leq m_{s_0} k_d, \tag{4.6}$$

where $k_d > 0$ is a known constant value denoting the upper bound for the absolute value.

Preliminaries

In this section, we begin with the review of some terminologies referred to the Lyapunov stability theory and the finite-time stability for nonlinear systems as following form [2, 3]:

$$\begin{cases} \dot{x} = f(x), \\ x(0) = x_0, \end{cases} \tag{4.7}$$

where $f : U_0 \to R^n$ is continuous on an open neighborhood U_0 of the origin that satisfy $f(0) = 0$ for all $t \geq 0$. Assume that the system (4.7) admits unique solutions in forward time for all initial conditions.

Definition 4.1 The trivial solution $x = 0$ is said to be a (locally) finite-time-stable equilibrium of system (4.7) if the solution exists for any initial data $x_0 \in U \subset R^n$ where $U \subseteq U_0$ is an open neighborhood of the origin. Moreover the following statements hold.

- Finite-time convergence: There exists a function $T_s : U \backslash \{0\} \to (0, \infty)$, which is called the settling time, such that, for every initial value $x_0 \in U \backslash \{0\}$, the solution of (4.7) denoted by $s_t(x_0)$ is defined with $s_t(x_0) \in U \backslash \{0\}$ for $t \in [0, T_s(x_0))$, and satisfies $\lim_{t \to T_s(x_0)} s_t(x_0) = 0$, and $s_t(x_0) = 0$, if $t \geq T_s(x_0)$.
- Lyapunov stability: For every open set U_ε satisfied with $0 \in U_\varepsilon \subseteq U$, there exists an open subset U_δ submitted to $0 \in U_\delta \subseteq U$ and such that for every $x_0 \in U_\delta \backslash \{0\}$, $s_t(x) \in U_\varepsilon$ for all $t \in [0, T(x_0))$.

The origin is said to be a globally finite-time stable equilibrium if it is a finite-time stable equilibrium with $U = U_0 = R^n$. Then the system is said to be finite-time convergent with respect to the origin.

Definition 4.2 Consider a nonlinear system as follows:

$$\dot{x} = f(x, u_f), \; x \in R^n, \; u_f \in R^m \tag{4.8}$$

where $f(0, 0) = 0$ and $f : R^{n \times m} \to R^n$ is globally defined. It is said that the origin $x = 0$ of (4.8) is finite-time stabilizable via continuous time-invariant state feedback if there is a continuous feedback law of the form $u_f = \mu(x)$ such that the origin of the closed-loop system $\dot{x} = f(x, \mu(x))$ is a (locally) finite-time stable equilibrium.

Remark 4.3 It is well known that the finite-time stability only exists in the systems of the form (4.7) with $f(x)$ non-Lipschitz. Thus the closed-loop system may be made nonsmooth by the control law $\mu(x)$. Therefore, the class of homogeneous functions, such as [4] , are often sought to synthesize the finite-time control laws. Thus, let us introduce the concept of homogeneity following the treatment of [5] for the subsequent analysis.

Definition 4.4 For fixed coordinates $(x_1, \ldots, x_n) \in R^n$ and real numbers $r_i > 0, i = 1, \ldots, n$,

- the dilation $\Delta_\varepsilon(x)$ is defined by $\Delta_\varepsilon(x) = (\varepsilon^{r_1} x_1, \ldots, \varepsilon^{r_n} x_n)$ for any $\varepsilon > 0$, where r_i is called the weights of the coordinates.
- a function $V \in (R^n, R)$ is said to be homogeneous of degree τ if there is a real number $\tau \in R$ such that $V(\Delta_\varepsilon(x)) = \varepsilon^\tau V(x_1, \ldots, x_n)$ for any $x \in R^n \backslash \{0\}, \varepsilon > 0, i = 1, \ldots, n$.
- a vector field $f \in (R^n, R^n)$ is said to be homogeneous of degree τ if there is a real number $\tau \in R$ such that $f_i(\Delta_\varepsilon(x)) = \varepsilon^{\tau + r_i} f_i(x)$ for any $x \in R^n \backslash \{0\}, \varepsilon > 0, i = 1, \ldots, n$.

System (4.7) is called homogeneous if its vector field f is homogeneous.

Lemma 4.5 *([2]) Suppose that system (4.7) is homogeneous of degree τ. If the origin is asymptotically stable and $\tau < 0$, then the origin of the system is finite-time stable.*

System Requirements

The system requirements can be summarized as follows.

- Due to the disturbances caused by irregular road, a firm uninterrupted contact between the tire and the road is an important preliminary for vehicle handling and is essentially related to ride safety. Thus, the tire is enabled to transfer the longitudinal and lateral forces to the vehicle, so that the brake inputs, steering and throttle can be controlled by the driver. Therefore, the dynamic tire load ($F_{\text{dyn}} = F_t + F_b$) has to be less than the static tire load and be bounded by

$$\max(\| F_{\text{dyn}} \|) \leq (m_s + m_u)g = F_{\text{stat}}, \tag{4.9}$$

  where g stands for the gravitational constant and F_{stat} is the static tire load.
- For the sake of avoid damaging vehicle components and generating more passenger discomfort, the active suspension controllers must be capable of preventing the suspension from hitting its travel limits. In addition, when the suspension is subject to road-induced vibrations, the standard deviation $\|z_s - z_u\|$ is considered for the analysis of the suspension deflection. Therefore, it is necessary to make sure that the suspension deflection should not exceed the allowable the maximum suspension deflection hard limit $z_{\max}$, which can be described as

$$\|z_s - z_u\| \leq z_{\max}. \tag{4.10}$$

Problem Statement

Synthesize a control strategy u which induces an ideal sliding-mode motion in the prescribed sliding-mode surface, such that the active suspension system (4.5) is said to solve a globally finite-time stable problem along the sliding-mode surface asymptotically, and meanwhile, stabilize the vertical motion of the closed-loop system in the presence of parametric uncertainty and uncertain nonlinearity and isolate the force transmitted to the passengers as well.

4.1.2 *Main Results*

Sliding Mode Dynamics Analysis

The sliding-mode technique provides an effective and robust method for controlling nonlinear time-varying systems having disturbances. Such a control method employs a discontinuous control law to drive the system toward a specified sliding surface and maintain the system's motion along the sliding surface in the state space. Thus, in this section, a chattering free sliding-mode control strategy is presented to ensure the stability of sliding mode dynamics, and this strategy is synthesized to drive the state trajectories of system (4.5) onto the prespecified sliding surface. In addition, as for active suspension systems, the main objective of the control is to reduce the effect of the road irregularities on the passengers and to insure the system safety during vehicle maneuvers.

For convenience, as in [6], set $\mathrm{sig}(y)^\alpha = \mathrm{sgn}(y)|y|^\alpha$ for $\alpha > 0$, where $|y|$ denotes the absolute value of real number y and $\mathrm{sgn}(\cdot)$ the sign function. Clearly, $\mathrm{sig}(y)^\alpha = |y|^\alpha$ if $\alpha = p_1/p_2$ where $p_i > 0$, $i = 1, 2$ are odd integers.

(1) Sliding surface
Firstly, the switching surface s is constructed for third-order systems as a combination of the state variables x_1, x_2 and $\dot{x}_2$ as follows:

$$s = \dot{x}_2 + c_2\mathrm{sig}(x_2)^{\alpha_2} + c_1\mathrm{sig}(x_1)^{\alpha_1}, \tag{4.11}$$

where c_i and α_i $(i = 1, 2)$ are constants. c_i are to be designed such that the polynomial $p^2 + c_2 p + c_1$, which corresponds to system (4.11), is Hurwitz, i.e., the eigenvalues of the polynomial are all in the left-half side of the complex plane. Thus, it is easy to imply that c_1 and c_2 are positive constants. α_i are positive values and satisfy the following condition as [4]:

$$\alpha_1 = \frac{\alpha}{2-\alpha}, \quad \alpha_2 = \alpha, \tag{4.12}$$

where $\alpha \in (0, 1)$.

When the state trajectories of the system (4.5) enter into the ideal sliding mode $s = 0$, the active suspension system will behave in an identical fashion, namely

$$\begin{cases} \dot{x}_1 = x_2, \\ \dot{x}_2 = -c_2\mathrm{sig}(x_2)^{\alpha_2} - c_1\mathrm{sig}(x_1)^{\alpha_1}. \end{cases} \tag{4.13}$$

Proposition 4.6 *The origin of (4.13) is a globally finite-time stable equilibrium. In addition, both the states x of (4.13) are bounded with any bounded initial condition for $x(0)$.*

Proof According to the description of system (4.7), we rewrite the system (4.13) in the following form

$$\dot{x} = f(x) = (f_1(x), f_2(x))^T, \ x = (x_1, x_2)^T, \tag{4.14}$$

where

$$\begin{cases} f_1(x_1, x_2) = x_2, \\ f_2(x_1, x_2) = -c_1 sig(x_1)^{\alpha_1} - c_2 sig(x_2)^{\alpha_2}. \end{cases} \tag{4.15}$$

Applying the condition (4.12) results in

$$\frac{2}{1+\alpha_1}\alpha_1 = \alpha_2. \tag{4.16}$$

Furthermore, it is seen that for any $\varepsilon > 0$

$$\begin{aligned} f_1(\varepsilon^{\frac{2}{1+\alpha_1}} x_1, \varepsilon x_2) =& \varepsilon x_2 = \varepsilon^{\tau + \frac{2}{1+\alpha_1}} x_2, \\ f_2(\varepsilon^{\frac{2}{1+\alpha_1}} x_1, \varepsilon x_2)) =& - c_1 \varepsilon^{\frac{2}{1+\alpha_1}\alpha_1} sig(x_1)^{\alpha} - c_2 \varepsilon^{\alpha_2} sig(x_2)^{\alpha_2} \\ =& \varepsilon^{\tau+1} f_2(x_1, x_2), \end{aligned}$$

where $\tau = \alpha_2 - 1 < 0$ and condition (4.16) are used. Therefore, based on the Definition 4.4, the vector field f is homogenerous of degree $\tau = \alpha_2 - 1 < 0$ with weights $(2/(1+\alpha_1), 1)$. That is to say, system (4.13) is homogeneous of negative degree τ with respect to $(2/(1+\alpha_1), 1)$.

Consider the continuously differentiable Lyapunov function candidate $W : R^2 \to R$ of the form given by $W(x_1, x_2) = \frac{1}{2}x_2^2 + \frac{c_1 |x_1|^{1+\alpha_1}}{1+\alpha_1}$. Then, a direct computation shows that $L_f W(x) = -c_2 |x_2|^{1+\alpha_2} \le 0$. Due to $L_f W(x)(x) \equiv 0$ together with (4.13), it is easy to imply $(x_1, x_2) \equiv 0$. Thus, by means of LaSalle's invariant set theorem, the equilibrium $(x_1, x_2) \equiv 0$ of system (4.13) is globally asymptotically stable.

From Theorem 6.2 of [7], there is a continuous, positive definite Lyapunov function $V : R^2 \to R$ such that $\nabla_x V$ is continuous on R^2. Moreover, V is homogeneous of degree $l > \max\{-\tau, \frac{2}{1+\alpha_1}, 1\}$, and $L_f V$ is homogeneous of degree $l + \tau$, both with the same weights $(\frac{2}{1+\alpha_1}, 1)$. We also known from theorem 2 of [5] that V is radially unbounded. By means of Lemma 4.2 of [7], there exists a $c > 0$ such that $L_f V(x) \le -c(V(x))^{\frac{l+\tau}{l}}$. Moreover, the settling-time estimation have been given in Theorem 4.2 of [8], it is shown that $T_{s_1}(x_0) \le (1/c(\frac{1-\alpha_2}{l}))(V(x_0))^{\frac{1-\alpha_2}{l}}$, for all x_0 in the neighborhood of origin. In this way, the conclusion follows from Lemma 4.5.

At the same time, with the result $L_f W(x) \leq 0$, it not hard to imply that $W(x) \leq W(x(0))$, where $W(x(0))$ is a bounded positive constant. Furthermore, it is easy to check that

$$|x_1| \leq \sqrt[1+\alpha_1]{\frac{1+\alpha_1}{c_1} W(x(0))}, \ |x_2| \leq \sqrt{2W(x(0))}. \tag{4.17}$$

Thus, x_1 and x_2 are bounded with the bounded initial condition. □

Remark 4.7 System (4.13), which denotes for the establishment of the ideal sliding-mode $s = 0$ for system (4.5), can converge to its equilibrium point $(x_1, x_2)^T = [0, 0]^T$ in finite-time, if c_1, c_2 are designed to ensure that the the polynomial $p^2 + c_2 p + c_1$ is Hurwitz and α_1, α_2 can be determined based on (4.12).

(2) Performance of the sliding motion
In this section, based on Lyapunov functions, we prove the finite-time stable of the closed-loop tracking error. Now, the main result is stated.

Theorem 4.8 *The active suspension system (4.5) will reach the sliding-mode surface $s = 0$ in finite-time and then converge to zero along $s = 0$ within finite-time, if the sliding-mode surface s is chosen as (4.11) and the control strategy is designed as follows:*

$$u = m_{s0}(u_{eq} + u_n), \tag{4.18}$$
$$u_{eq} = -\phi(x,t)/m_{s0} - c_2 sig(x_2)^{\alpha_2} - c_1 sig(x_1)^{\alpha_1}, \tag{4.19}$$
$$\dot{u}_n + T_f u_n = v, \tag{4.20}$$
$$v = -(k_d + k_T + \eta) sgn(s), \tag{4.21}$$

where $u_n(0) = 0$; c_i and α_i $(i = 1, 2)$ are all constants, as defined in (4.13); η is a positive constant; k_d is a constant defined in (4.6); two constants, $T_f \geq 0$ and k_T are selected to satisfy the following condition:

$$k_T \geq T_f l_d. \tag{4.22}$$

Proof By combining system dynamic (4.5) and (4.11), the sliding–mode manifold can be expressed in another form as

$$\begin{aligned} m_{s0} s &= m_{s0}\dot{x}_2 + m_{s0}(c_2 \mathrm{sig}(x_2)^{\alpha_2} + c_1 \mathrm{sig}(x_1)^{\alpha_1}) \\ &= \phi(x,t) + u + F_\Delta + m_{s0}(c_2 \mathrm{sig}(x_2)^{\alpha_2} + c_1 \mathrm{sgn}(x_1)^{\alpha_1}). \end{aligned}$$

Substituting the control (4.18) into above equation gives:

$$\begin{aligned} m_{s0} s = &\ \phi(x,t) + F_\Delta + m_{s0}(u_{eq} + u_n) \\ &+ m_{s0}(c_2 \mathrm{sig}(x_2)^{\alpha_2} + c_1 \mathrm{sig}(x_1)^{\alpha_1}). \end{aligned} \tag{4.23}$$

The following equation can be derived from (4.19) and (4.23):

$$m_{s_0}s = F_\Delta + m_{s_0}u_n. \tag{4.24}$$

If the sliding-mode manifold $s > 0$, then it is not hard to get

$$0 < m_{s_0}s = F_\Delta + m_{s_0}u_n \le m_{s_0}l_d + m_{s_0}u_n, \tag{4.25}$$

that is, $l_d + u_n > 0$. On both sides of the inequality is multiplied by T_f, one obtains that $T_f l_d + T u_n > 0$. Based on (4.22), the following relationships under the condition $s > 0$ can be obtained: $k_{T_u} + T_f u_n > 0$, and $T_f u_n s + k_{T_u}|s| > 0$.

Similarly, when the sliding-mode manifold $s < 0$, then we have $0 > s = m_{s_0}s = F_\Delta + m_{s_0}u_n \ge -m_{s_0}l_d + m_{s_0}u_n$, which we can further get $-l_d + u_n < 0$. Following a similar procedure as $s > 0$, the following inequality is obtained: $-k_{T_u} + T_f u_n < 0$. According to condition $s < 0$, we have $T_f u_n s + k_{T_u}|s| > 0$.

Finally, once the sliding-mode manifold $s = 0$, one has $T_f u_n s + k_{T_u}|s| = 0$.

The aforementioned analysis yields to the following inequality kept forever:

$$T_f u_n s + k_{T_u}|s| \ge 0. \tag{4.26}$$

Define a positive Lyapunov function as following: $V = \frac{1}{2}m_{s_0}s^2$. For sliding-mode surface (4.11), its derivative with respect to time t along system (4.5) can be obtained from (4.24) as follows:

$$\begin{aligned} m_{s_0}\dot{s} &= \dot{F}_\Delta + m_{s_0}\dot{u}_n = \dot{F}_\Delta + m_{s_0}\dot{u}_n + m_{s_0}T_f u_n - m_{s_0}T_f u_n \\ &= \dot{F}_\Delta + m_{s_0}v - m_{s_0}T_f u_n. \end{aligned} \tag{4.27}$$

Substituting (4.21) into above equation, the sliding-mode surface dynamic can be expressed as

$$m_{s_0}\dot{s} = \dot{F}_\Delta - m_{s_0}(k_d + k_T + \eta)\text{sgn}(s) - m_{s_0}T_f u_n, \tag{4.28}$$

By multiplying s on both sides of (4.28), this study gets that

$$\begin{aligned} m_{s_0}s\dot{s} =& \dot{F}_\Delta s - m_{s_0}(k_d + k_T + \eta)|s| - m_{s_0}T_f u_n s \\ =& (\dot{F}_\Delta s - m_{s_0}k_d|s|) + m_{s_0}(-T_f u_n s - k_T|s|) - m_{s_0}\eta|s|. \end{aligned}$$

From (4.6), (4.26) and above equation, the following inequality is obtained:

$$\dot{V} = m_{s_0}s\dot{s} \le -m_{s_0}\eta|s| = -\sqrt{2m_{s_0}}\eta V^{\frac{1}{2}} < 0 \text{ for } |s| \ne 0. \tag{4.29}$$

Since $\sqrt{2m_{s_0}}\eta > 0$, expression (4.29) proves that TSM $s = 0$ is attainable in finite time less or equal to $T_{s_2} = \frac{2V(0)^{\frac{1}{2}}}{\sqrt{2m_{s_0}}\eta}$. Consequently, from Proposition 4.6 and

Definition 4.2, it follows that the origin system (4.5) behaves in an identical fashion, as shown in (4.13), i.e., the system will converge to zero in a finite-time along $s = 0$, where the settling-time estimation given by $T_s \leq T_{s_1} + T_{s_2}$ with the initial state of the system.

This completes the proof. □

Remark 4.9 The control input is expressed as the sum of two terms. The first one, called the equivalent control, is chosen ignoring the nominal plant parameters (F_Δ), so as to make $\dot{s} = 0$ when $s = 0$. It is given by (4.19). The second term is chosen to tackle the uncertainties in the system and to introduce a reaching law, which is different from the design method as usual and can be effectively used to avoid control chattering problem. As stated in Theorem 4.8, the control signal (4.20) is equivalent to a low-pass filter, where $v(t)$ is the input and $u_n(t)$ is the output of the filter. The transfer function corresponding to the filter (4.20) is:

$$\frac{u_n(s)}{v(s)} = \frac{1}{s + T_f}, \tag{4.30}$$

where the bandwidth of the low pass filter is equal to the parameter T_f. Although the control law $v(t)$ in (4.21) is discontinuity caused by the switch function sgn(s), $u_n(t)$ in (4.18) is softened to be a smooth signal by (4.20).

As a special case, when the bandwidth parameter $T_f = 0$, then the control laws (4.20) and (4.21) can be rewritten as following:

$$\dot{u}_n = v, \ v = -(k_d + \eta)\text{sgn}(s). \tag{4.31}$$

Noting that when the control law in (4.20) and (4.21) are substituted for (4.31), it is easy to get the control law u in (4.18) is still continuous and the Theorem 4.8 holds as well. In this case, the proof process of the Theorem 4.8 do not require the condition (4.22). Obviously, (4.31) is a pure integrator form compared with the the low pass filter (4.20), which is difficult for hardware implementation in practical applications.

Remark 4.10 It is noteworthy that, with preventing differentiation terms $c_i \text{sig}(x_i)^{\alpha_i}$ in the sliding-mode surface (4.11) from deriving the proposed control laws, the ideal sliding–mode surface, $s = 0$, discussed above is nonsingular. Thus, the singularity in sliding mode control can be avoided.

Remark 4.11 In the limiting case where α_1 approaches 1 and the second term of (4.18) is neglected, then the controller becomes the conventional proportional-derivative (PD) controller, which is a smooth control law, and cannot achieve finite-time stability as described in Remark 4.3.

Remark 4.12 It should be mentioned that in (4.18)–(4.21), all variables are available except s on account of $\dot{x}_2$ is not available in (4.11). For calculating sgn(s) in (4.21), the following function $h(t)$ has to be considered:

$$h(t) = \int_0^t s(t)\mathrm{d}t = x_2 + \int_0^t (c_2 \mathrm{sig}(x_2)^{\alpha_2} + c_1 \mathrm{sig}(x_1)^{\alpha_1})\mathrm{d}t. \tag{4.32}$$

sgn(s) can be calculated through the following equation $\mathrm{sgn}(s) = \mathrm{sgn}(h(t) - h(t - \tau_0))$, where τ_0 is a time delay constant and can be chosen as a fundamental sample time because $s(t) = \lim_{\tau_0 \to 0}(h(t) - h(t - \tau_0))/\tau_0$. In this way, the accurate value of s do not need to be known except its sign, sgn(s), that is to say, we only need to know whether h increases or decreases. It should be noted that to obtain sgn(s) is much easier than to obtain the accurate value of s.

Zero Dynamics

The above design yields to a 2nd order error dynamic, while the original system is a 4th order system. So the zero dynamics consist of two states. Therefore, the zero dynamic subsystem is obtained by setting $x_1 = 0$ together with the corresponding derivative $\dot{x}_1 = x_2 = 0$. As such, the 2nd equation of (4.5) can be described by:

$$m_{s_0}\dot{x}_2 = \phi(x, t) + u + F_\Delta = 0, \tag{4.33}$$

which, in essence, is an algebraic equation. Hence, the control input u can be obtained as

$$u = -\phi(x, t) - F_\Delta, \tag{4.34}$$

Furthermore, substituting u in (4.34) into the bottom equation of (4.5) to replace the one in x_4, the zero dynamics equation is obtained by

$$\dot{x} = Ax + w, \tag{4.35}$$

where $x = [x_3, x_4]^T$, $A = \left[0, 1; -\frac{k_f}{m_u}, -\frac{b_f}{m_u}\right]$, $w = \left[0, \frac{k_f}{m_u}z_r + \frac{b_f}{m_u}\dot{z}_r + \frac{1}{m_u}F_\Delta\right]^T$. Now, considering a positive definite function $V_0 = x^T x$, then from (4.35) we get

$$\dot{V}_0 = x^T(A^T + A)x + 2x^T w. \tag{4.36}$$

Clearly, it is easy to verify that the matrix A has eigenvalues with negative real parts. Hence, we can have $A^T + A < 0$. Noting that

$$2x^T w \leq \frac{1}{\xi}x^T x + \xi w^T w, \tag{4.37}$$

where ξ is a tuning positive value, and assuming that the disturbance $\xi w^T w$ is bounded by κ_2, then (4.36) can be equivalently expressed as

$$\dot{V}_0 \leq \left[-\lambda_{\min}(A^T + A) + \frac{1}{\xi}\right]x^T x + \kappa_2. \tag{4.38}$$

Based on properly chosen tuning value ξ, we can guarantee

$$-\lambda_{\min}(A^T+A)+\frac{1}{\xi}=-\kappa_1. \tag{4.39}$$

It is easily shown from (4.39) we can conclude that the Lyapunov function is bounded by

$$\begin{aligned} V_0(t) &\le V_0(0)e^{-\kappa_1 t}+\frac{\kappa_2}{\kappa_1}(1-e^{-\kappa_1 t}) \\ &\le \max\{V_0(0),\frac{\kappa_2}{\kappa_1}\}=q_2, \end{aligned} \tag{4.40}$$

which results in that $|x_j|\le\sqrt{q_2}$, $j=3,4$.

From the above proof, we can see all the four states are constrained within their known bounds, and therefore, the performance constraints can be guaranteed by adjusting the bounds of the states as follows:

$$|x_1-x_3|\le \sqrt[1+\alpha_1]{\frac{1+\alpha_1}{c_1}W(x(0))}+\sqrt{q_2}, \tag{4.41}$$

if $\sqrt[1+\alpha_1]{\frac{1+\alpha_1}{c_1}W(x(0))}+\sqrt{q_2}\le z_{\max}$, then the suspension space will be constrained within its range. Similarly, we have

$$|F_t+F_b|\le k_f(\sqrt{q_2}+d_{w1})+b_f(\sqrt{q_2}+d_{w2}), \tag{4.42}$$

where d_{w1} and d_{w2} are the upper bounds of the disturbances z_r and $\dot{z}_r$. If we adjust the initial values and tuning parameters to meet $k_f(\sqrt{q_2}+d_{w1})+b_f(\sqrt{q_2}+d_{w2})\le (m_s+m_u)g$, then we can guarantee the constrained condition (4.9).

4.1.3 Comparative Experimental Results

In this section, it is intended to implement a controller for an actual hardware setup of an active suspension system in the laboratory aimed to validate the proposed control and reject some external disturbances. The experimental setup consisting of a bench-scale model to emulate a quarter-car model is illustrated in Fig. 4.2, whose model parameters are listed in Table 4.1.

This active suspension system consists of three masses, or plates, which can independently move in the vertical direction from each other. The bottom plate is driven by a brushed servo motor connected to a lead screw and cable transmission system, which is used to generate different road profiles and the perturbation of the system. The middle plate is linked by a spring and a damper to the bottom plate. There

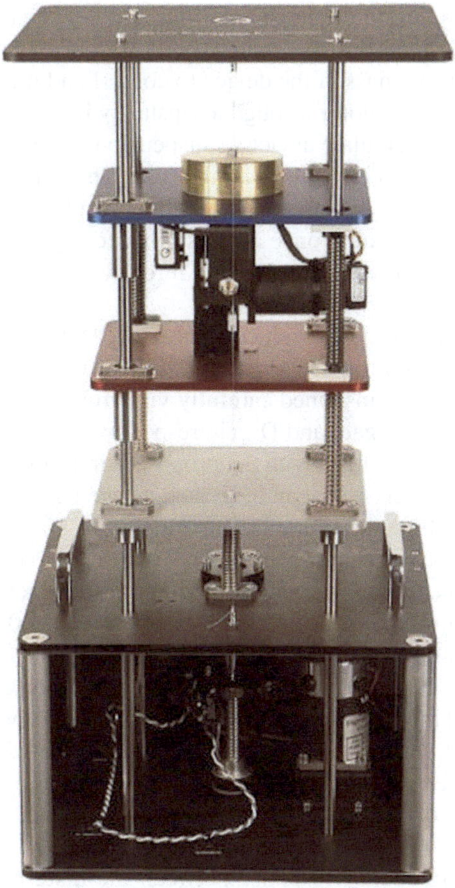

Fig. 4.2 The structure of quarter-car active suspension setup

Table 4.1 The model parameters of active suspensions setup

Parameter	Value	Units	Parameter	Value	Units
m_s	2.45	kg	k_f	2500	N/m
m_u	1	kg	b_f	1000	Ns/m
k_s	900	N/m	b_e	8	Ns/m
k_{s_n}	10	N/m	b_c	7	Ns/m

is damping due to friction in the linear bearings and between the springs and their guide. The top plate represents the vehicle body supported above the suspension, also known as the sprung mass of the device to control and it is linked to the middle plate by a high-quality DC motor through a capstan, which is known as the actuator of the control system, to emulate an active suspension system that can dynamically compensate for the motions introduced by the road. The aim of the control problem is to minimize the energy of the acceleration of the first plate, by regulation the force provided by the second actuator, once any disturbance is given.

For subsequent comparison, the following three systems are given respectively:

(1) Passive suspension systems.

(2) Active suspension systems with conventional PD controller. This is the traditional position tracking controller. The PD controller is implemented in the real-time control software, and the gains tuned carefully via error and try method are $k_p = 6$, $k_i = 5$, which denote the P-gain and D-gain respectively.

(3) Active suspension systems with the proposed finite-time controller. Since the robust term $k_d + k_T + \eta$ in (4.21) can be treated as one parameter η, thus, only gain η needs to be tuned during operating. Thus, there are six parameters need to be tuned. The control parameters are chosen as $c_1 = 6$, $c_2 = 5$, $\alpha_1 = 9/23$, $\alpha_2 = 9/16$, $T_f = 0.1$, $\eta = 10$. The nominal crude estimation of m_s is taken as $m_{s_0} = 2.2\,\text{kg}$.

The proposed controller was tested by 3 Hz sine wave of 0.2 cm amplitude, that is $z_r = 0.002\sin(6\pi t)$.

Figure 4.3 shows a comparison between passive (red line), conventional PD controller (blue line) and the proposed controller (black line) for vertical displacements z_s, and it can be seen that our proposed controller can stabilize the vertical motion best in spite of the presence of extern disturbance.

The improvement of the closed-loop system with the resulting controller with respect to the passive response can be evaluated in Fig. 4.4, where the magnitude of the experimental frequency response between the disturbance and the vertical displacement z_s is illustrated. Notice that, despite in conventional PD controller approach the choice of control parameters is simple, the constraint in the controller structure makes the final performance worse than the finite-time approach. In particular, it can be noticed that the frequency corresponding to the maximum amplification factor, i.e, 3 Hz, the finite-time approach leads to an additional disturbance reduction of almost 10^{-2} order of magnitude.

In active suspension control, it is widely accepted that ride comfort is closely related to the body acceleration. Here, the time-domain responses of the body vertical accelerations for the suspension are illustrated in Fig. 4.5, which shows a comparison between the aforementioned three systems of the vertical acceleration $\ddot{z}_s$. It is observed from these figures that our proposed approach improves suspension performances in terms of peak response values compared with the other two systems. These results confirm the efficiency of our designed controllers.

Moreover, the limitations of the suspension space should be taken into account, which means that the suspension working space must be preserved. The suspension stroke $z_s - z_u$ for passive and active suspensions is illustrated in Fig. 4.6, which can be observed that the controlled suspension spaces all fall into the acceptable ranges

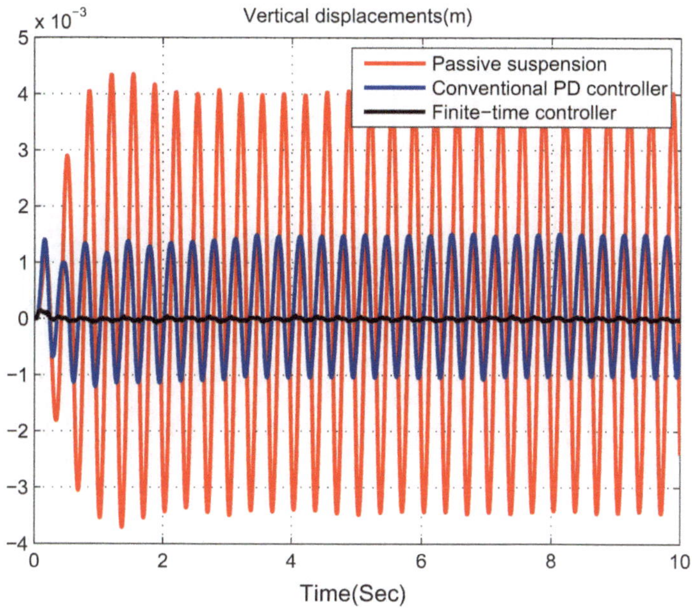

Fig. 4.3 Displacement responses of the vertical motion

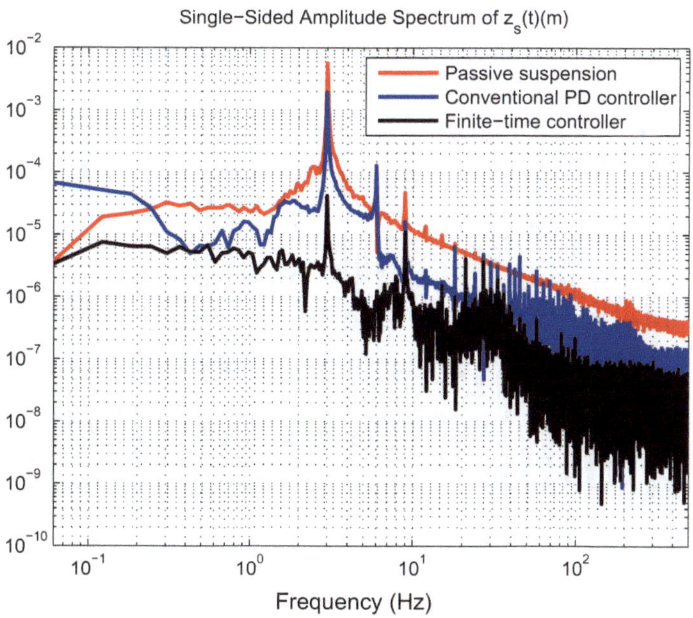

Fig. 4.4 Frequency responses of the vertical motion z_s

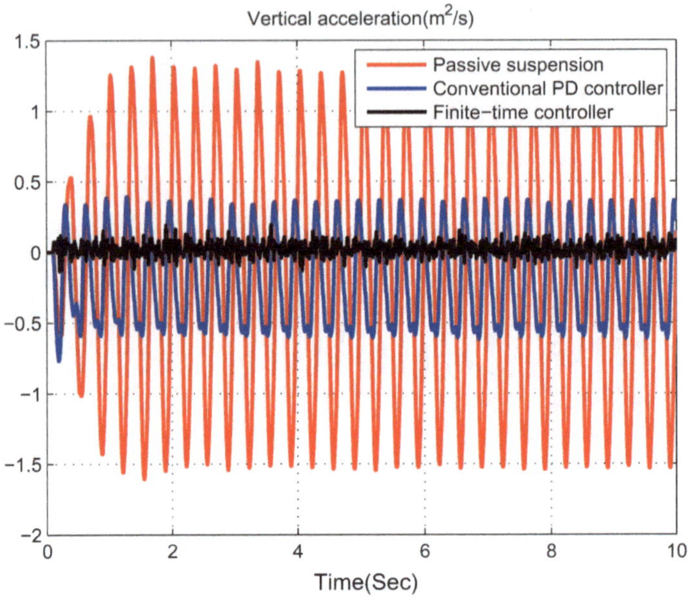

Fig. 4.5 Vertical acceleration

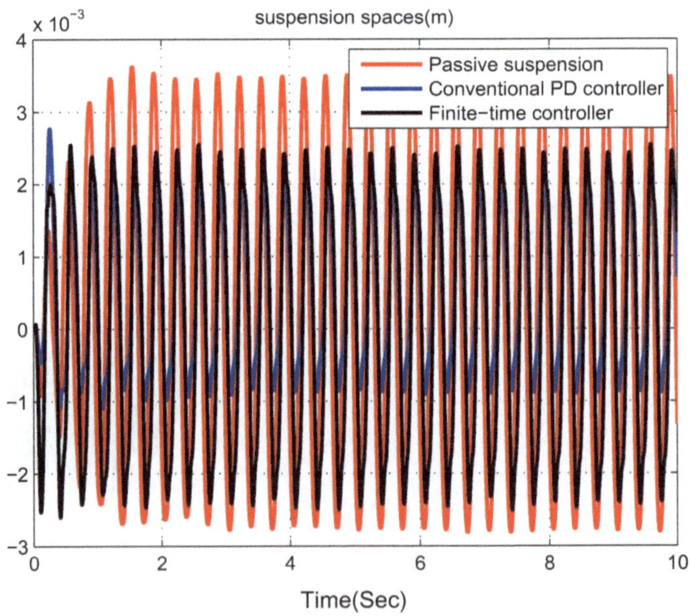

Fig. 4.6 Suspension spaces

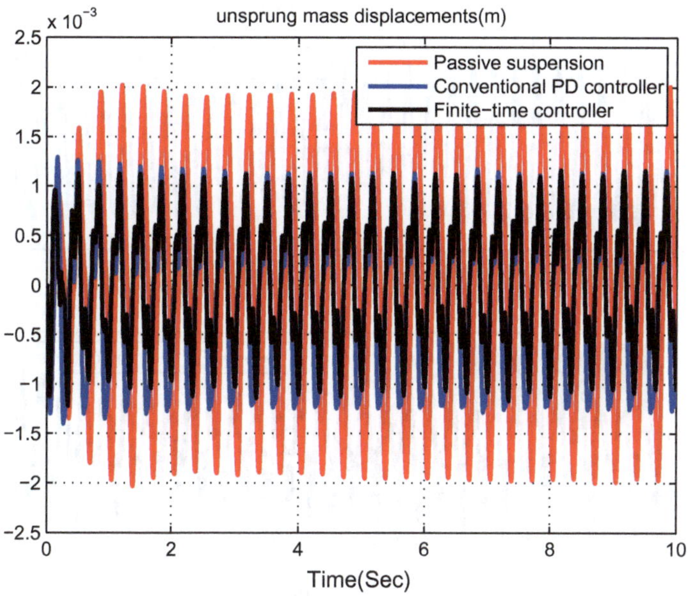

Fig. 4.7 The responses of unsprung mass displacement $z_u - z_r$

Table 4.2 RMS of system states

States ($\times 10^{-4}$)	Passive	PD (percentage)	FC (percentage)
z_s	26	9.3806 ($\uparrow$ 63.92%)	0.25368 ($\uparrow$ 99.02%)
$\ddot{z}_s$	9463	3924 ($\uparrow$ 58.53%)	467 ($\uparrow$ 95.07%)
$z_s - z_u$	23	12 ($\uparrow$ 47.83%)	18 ($\uparrow$ 21.74%)
z_u	13	7.9749 ($\uparrow$ 38.65%)	6.6686 ($\uparrow$ 48.70%)

$z_{\max} = 0.038$ m, although the proposed controller may need bigger suspension space than the conventional PD controller. That is because an active suspension system is more elastic and efficient, which can provide more handling capability and ride quality by both add and dissipate energy from the system. Thus, a comfortoriented suspension calls for a low damping and a large stroke of the chassis mass to provide sufficient isolation. The vibration isolation properties of a suspension can be enhanced by softer primary springs, which allows for too much movement.

Figure 4.7 shows the stability of the zero dynamic systems, from which we can see that the state of the zero dynamic systems are stable. Figure 4.8 shows the trajectories of the actuator forces.

In order to evaluate the suspension system performance and the improvement in ride comfort, the root mean square (RMS) values of the vehicle body are exploited to demonstrate the effectiveness of the proposed controller design method. The RMS values are strictly related to the ride comfort of passengers, especially in the frequency

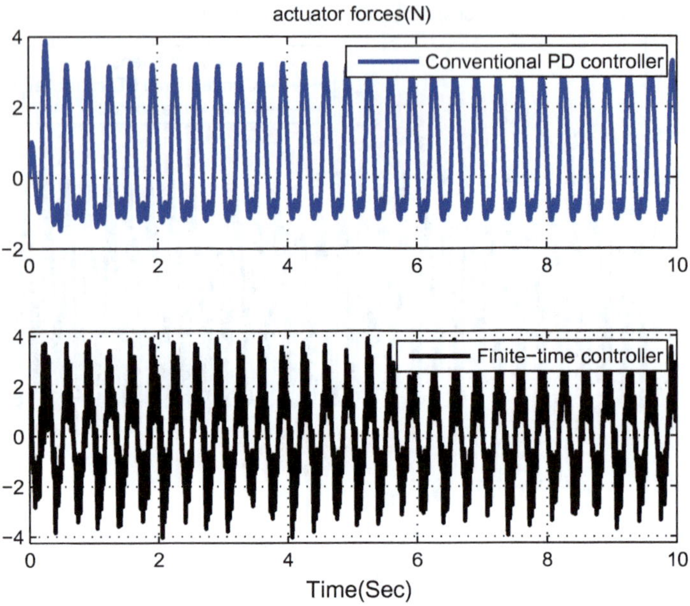

Fig. 4.8 Control input

range 2–8 Hz as pointed out in Section I, which are often used to quantify the amount of acceleration transmitted to the vehicle body. The RMS value of variable $x(t)$ is calculated as $\text{RMS}_x = \sqrt{(1/T)\int_0^T x^T(t)x(t)dt}$. In this study, $T = 10$ s is chosen to calculate and presented the RMS values.

These results are quantified using RMS. Table 4.2 shows a comparison between passive, PD and finite-time controller using RMS, and this is for the chassis states z_s, the vertical body acceleration $\ddot{z}_s$, the suspension strokes $z_s - z_u$, and the tire deflection z_u. It also shows the percentage of motion increase/decrease in active suspension compared with those of passive ones. It is clear that active suspension improves ride comfort by reducing the effect of road perturbations on the chassis. As can be seen, this improvement is high for the heave. Suspension strokes are bigger in active suspension in order to compensate for road concavities and convexities, as previously explained. In addition, active suspension reduces tire deflections, which results in increasing system security and road handling.

4.1.4 Conclusion

In this section, a new class of continuous TSM control strategy for the trajectory tracking of active suspensions with the finite-time stabilization problem has been studied. The new form of TSM can be used to design the controller not only because of its finite-time convergence to a given equilibrium, but also the continuous TSM control laws to drive system states convergence to a corresponding sliding surface in finite-time. Thus, the research has offered an alternative approach for improving the design of the suspension controller, and also solved the singularity and chattering problems in TSM systems. Also experiment results show that the control law can achieve faster response than conventional PD control law. This virtue may be attributed to the extra parameters in the control law.

4.2 Constrained Adaptive Backstepping Control for Uncertain Nonlinear Active Suspension Systems

This section proposes an adaptive backstepping control strategy for vehicle active suspensions with hard constraints. An adaptive backstepping controller is designed to stabilize the attitude of vehicle and meanwhile improve ride comfort in the presence of parameter uncertainties, where suspension spaces, dynamic tire loads and actuator saturations are considered as time-domain constraints. In addition to spring nonlinearity, the piece-wise linear behavior of the damper, which has different damping rates for compression and extension movements, is taken into consideration to form the basis of accurate control. Furthermore, a reference trajectory is planned to keep the vertical and pitch motions of car body to stabilize in pre-determined time, which helps adjust accelerations accordingly to high or low levels for improving ride comfort. Finally, a design example is shown to illustrate the effectiveness of the proposed control law.

4.2.1 *Problem Formulation*

Nonlinear Half-Car Model

In this section, the nonlinear half-car model is the second model in Sect. 1.2 and friction forces of suspension components are neglected. This model has been used extensively in the literature and captures many important characteristics of vertical and pitch motions. Here, the effect of lateral motion is neglected. The ideal dynamic equations of the sprung and unsprung masses are given by:

$$\begin{aligned} M\ddot{z}_c + F_{df} + F_{dr} + F_{sf} + F_{sr} &= u_z, \\ I\ddot{\varphi} + a(F_{df} + F_{sf}) - b(F_{dr} + F_{sr}) &= u_\varphi, \\ m_f\ddot{z}_1 - F_{sf} - F_{df} + F_{tf} + F_{bf} &= -u_1, \\ m_r\ddot{z}_2 - F_{sr} - F_{dr} + F_{tr} + F_{br} &= -u_2, \end{aligned} \tag{4.43}$$

where $u_z = u_1 + u_2$ and $u_\varphi = au_1 - bu_2$. The forces produced by the non-linear stiffening spring, the piece-wise linear damper and the tire obey:

$$F_{sf} = k_{f1}\Delta y_f + k_{nf1}\Delta y_f^3, \ F_{sr} = k_{r1}\Delta y_r + k_{nr1}\Delta y_r^3, \tag{4.44}$$

$$F_{df} = \begin{cases} b_{e1}\Delta\dot{y}_f, \\ b_{c1}\Delta\dot{y}_f, \end{cases} F_{dr} = \begin{cases} b_{e2}\Delta\dot{y}_r, \\ b_{c2}\Delta\dot{y}_r, \end{cases} \tag{4.45}$$

$$F_{tf} = k_{f2}(z_1 - z_{o1}), \ F_{tr} = k_{r2}(z_2 - z_{o2}), \tag{4.46}$$

$$F_{bf} = b_{f2}(\dot{z}_1 - \dot{z}_{o1}), \ F_{br} = b_{r2}(\dot{z}_2 - \dot{z}_{o2}), \tag{4.47}$$

where k_{f1}, k_{r1} and k_{nf1}, k_{nr1} are the stiffness coefficients of the linear and cubic terms; b_{ei} and b_{ci} $(i = 1, 2)$ are the damping coefficient for the extension and compression movements; $k_{f2}, k_{r2}, b_{f2}, b_{r2}$ are the stiffness and damping coefficients of the tires. Δy_f and Δy_r stand for the front and rear suspension spaces:

$$\Delta y_f = z_c + a\sin\varphi - z_1, \ \Delta y_r = z_c - b\sin\varphi - z_2. \tag{4.48}$$

Define the state variables $x_1 = z_c$, $x_2 = \dot{z}_c$, $x_3 = \varphi$, $x_4 = \dot{\varphi}$, $x_5 = z_1$, $x_6 = \dot{z}_1$, $x_7 = z_2$, $x_8 = \dot{z}_2$, and then the dynamic equations are rewritten in the following state-space form:

$$\begin{aligned}
\dot{x}_1 &= x_2, \\
\dot{x}_2 &= \frac{1}{M}(-F_{df} - F_{dr} - F_{sf} - F_{sr} + u_z), \\
\dot{x}_3 &= x_4, \\
\dot{x}_4 &= \frac{1}{I}(-a(F_{df} + F_{sf}) + b(F_{dr} + F_{sr}) + u_\varphi), \\
\dot{x}_5 &= x_6, \\
\dot{x}_6 &= \frac{1}{m_f}(F_{sf} + F_{df} - F_{tf} - F_{bf} - u_1), \\
\dot{x}_7 &= x_8, \\
\dot{x}_8 &= \frac{1}{m_r}(F_{sr} + F_{dr} - F_{tr} - F_{br} - u_2).
\end{aligned} \tag{4.49}$$

It is to be noted that with a change in the number of passengers or the payload, the vehicle load will easily vary and this will accordingly change the vehicle mass M and the moment of inertia I. In literature [9] can see numerous works that research into uncertain systems, some examples being [10–12]. In this section, we assume that $M_{\min} \le M \le M_{\max}$, $I_{\min} \le I \le I_{\max}$.

Problem Statement

For active suspension systems, the performance requirements include the following aspects.

1. Ride comfort: for active suspensions, the main task is to design a controller which can succeed in stabilizing the vertical and pitch motion of the car body and isolating the force transmitted to the passengers as well.
2. Good road holding: the dynamic tire load should not exceed the static ones for both of the front and rear wheels, i.e.

$$\left|D_f\right| = \left|F_{tf} + F_{bf}\right| < F_f, \quad |D_r| = |F_{tr} + F_{br}| < F_r, \tag{4.50}$$

where the static tire loads F_f and F_r are computed by

$$\begin{aligned} F_f + F_r &= (M + m_f + m_r)g, \\ F_f(a+b) &= Mgb + m_f g(a+b). \end{aligned} \tag{4.51}$$

3. Suspension space limits: because of mechanical structure, the suspension spaces should not exceed the allowable maximums, which can be described as

$$\left|\Delta y_f\right| \le \Delta y_{f\max}, \quad |\Delta y_r| \le \Delta y_{r\max}. \tag{4.52}$$

4. Actuator saturation: all actuators of physical devices are subject to amplitude saturation.

$$|u_i(t)| \le u_{i\max}, \quad i = 1, 2. \tag{4.53}$$

Based on the above statements, this study tries to deal with the following constrained adaptive control problem:

Problem 4.13 For the active suspension systems, synthesize adaptive control inputs u_i $(i = 1, 2)$ to stabilize the heave and pitch motions of closed-loop systems in the presence of parametric uncertainties, and meanwhile the essential performance requirements (ride safety conditions in (4.50), suspension space limits in (4.52) and actuator saturations in (4.53)) are guaranteed.

4.2.2 Control Law Synthesis

Adaptive Backstepping Controller Design

First, the control function u_z for the heave motion is designed to keep the tracking error $e_1 = x_1 - x_{1r}$ to converge to zero. The governing equations for the heave motion are:

$$\begin{aligned} \dot{x}_1 &= x_2, \\ \dot{x}_2 &= \theta_1(-F_z + u_z), \end{aligned} \tag{4.54}$$

where $F_z = F_{df} + F_{dr} + F_{sf} + F_{sr}$, and $\theta_1 = \frac{1}{M} \in [\theta_{1\min}\ \theta_{1\max}]$ with $\theta_{1\min} = \frac{1}{M_{\max}}$, $\theta_{1\max} = \frac{1}{M_{\min}}$.

Step 1: Design desired virtual control x_2^d, such that the tracking error e_1 is guaranteed to converge to zero asymptotically.

Starting with the equation of tracking error $e_1 = x_1 - x_{1r}$, we have $\dot{e}_1 = x_2 - \dot{x}_{1r}$, and let $e_2 = x_2 - x_2^d$. If we select this virtual control $x_2^d = \dot{x}_{1r} - k_1 \tanh(e_1)$, where k_1 is a positive value, then after considering a Lyapunov functional candidate

$$V_1 = \frac{1}{2}e_1^2, \tag{4.55}$$

the time derivative of V_1 becomes

$$\dot{V}_1 = e_1 e_2 - k_1 e_1 \tanh(e_1). \tag{4.56}$$

Clearly, if $e_2 = 0$, then $\dot{V}_1 = -k_1 e_1 \tanh(e_1) \leq 0$ and e_1 is guaranteed to converge to zero asymptotically.

Step 2: Synthesize an adaptive control law for u_z, so that the error e_2 converges to zero in the presence of unknown parameter θ_1.

Differentiating the error dynamics e_2 results in

$$\dot{e}_2 = \theta_1(-F_z + u_z) - \ddot{x}_{1r} + k_1(1 - \tanh^2(e_1))\dot{e}_1.$$

Choose the adaptive controller u_z as

$$u_z = \frac{1}{\hat{\theta}_1}(\ddot{x}_{1r} - k_1(1 - \tanh^2(e_1))\dot{e}_1 - k_2 \tanh(e_2) - e_1) + F_z.$$

The adaptation law is chosen as the projection type with the following form [13, 14]:

$$\dot{\hat{\theta}}_1 = Proj_{\hat{\theta}_1}(r_1\tau_1) = \begin{cases} 0, \text{ if } \hat{\theta}_1 = \theta_{1\max} \text{ and } r\tau > 0, \\ 0, \text{ if } \hat{\theta}_1 = \theta_{1\min} \text{ and } r\tau < 0, \\ r_1\tau_1, \text{ otherwise,} \end{cases}$$

where $r_1 > 0$ is a tunable gain and $\tau_1 = (-F_z + u_z)e_2$.

Choose a Lyapunov functional candidate

$$V_2 = V_1 + \frac{1}{2}e_2^2 + \frac{1}{2r_1}\tilde{\theta}_1^2. \tag{4.57}$$

Taking time derivative gives

$$\dot{V}_2 \leq -k_1 e_1 \tanh(e_1) - k_2 e_2 \tanh(e_2) \leq 0.$$

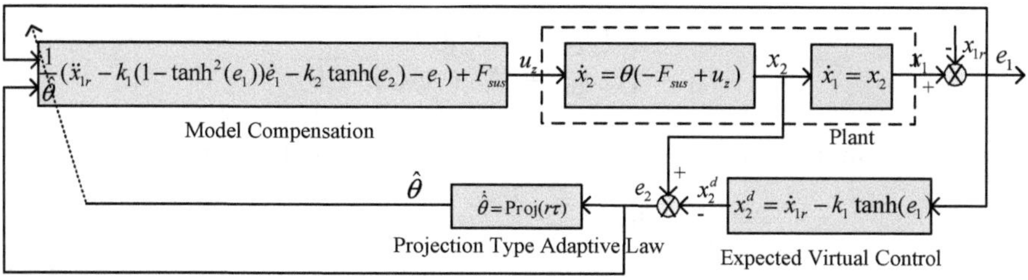

Fig. 4.9 Structure diagram of the vertical motion subsystem

Integrating both sides of inequality $\dot{V}_2 \le 0$ from 0 to t results in

$$V_2(t) = \int_0^t \dot{V}_2 d\tau + V_2(0) \le V_2(0), \tag{4.58}$$

which implies

$$|e_1| \le \sqrt{2V_2(0)}, \ \ |e_2| \le \sqrt{2V_2(0)}. \tag{4.59}$$

Equation (4.59) further leads to

$$\begin{aligned} |x_1| &\le |x_{1r}| + \sqrt{2V_2(0)} \le \|x_{1r}\|_\infty + \sqrt{2V_2(0)}, \\ |x_2| &\le \|\dot{x}_{1r}\|_\infty + (k_1 + 1)\sqrt{2V_2(0)}. \end{aligned} \tag{4.60}$$

From (4.60), it is true that $-F_z + u_z \in L_\infty$. Therefore, $\dot{e}_2 \in L_\infty$ and thus, $\ddot{V}_2$ is bounded. Therefore, $\dot{V}_2$ is uniformly continuous. By using Lyapunov-like lemma of [15], we have $\dot{V}_2 \to 0$ as $t \to \infty$, and then $e_1 \to 0$, $e_2 \to 0$, which means that the tracking errors e_1, e_2 converge to zero asymptotically. The structure diagram of the vertical motion subsystem is shown in Fig. 4.9.

Following a similar procedure, the resultant control function u_φ for the pitch motion can be obtained as:

$$\begin{aligned} e_3 &= x_3 - x_{3r}, \\ e_4 &= x_4 - x_4^d, \\ x_4^d &= \dot{x}_{3r} - k_3 \tanh(e_3), \\ u_\varphi &= \frac{1}{\hat{\theta}_2}(\ddot{x}_{3r} - k_3(1 - \tanh^2(e_3))\dot{e}_3 - k_4 \tanh(e_4) - e_3) + F_\varphi, \\ \dot{\hat{\theta}}_2 &= Proj_{\hat{\theta}_2}(r_2 \tau_2), \end{aligned}$$

where $F_\varphi = a(F_{df} + F_{sf}) - b(F_{dr} + F_{sr})$, $r_2 > 0$ is a tunable gain, $\tau_2 = (-F_\varphi + u_\varphi)e_4$, and k_3, k_4 are tuning gains. Furthermore, we have

$$|x_3| \le \|x_{3r}\|_\infty + \sqrt{2V_4(0)},$$
$$|x_4| \le \|\dot{x}_{3r}\|_\infty + (k_3+1)\sqrt{2V_4(0)}.$$

After obtaining u_z and u_φ, we can calculate the real inputs u_1 and u_2 as

$$u_1 = \frac{bu_z + u_\varphi}{a+b}, \quad u_2 = \frac{au_z - u_\varphi}{a+b}. \tag{4.61}$$

Zero Dynamics

The adaptive backstepping design yields to a 4th order error dynamic, while the original system is an 8th order system. So the zero dynamics consists of four states. To find it, we set $e_1 = e_3 = 0$. Hence, we obtain:

$$\begin{aligned} u_z &= M\ddot{x}_{1r} + F_z, \\ u_\varphi &= I\ddot{x}_{3r} + F_\varphi. \end{aligned} \tag{4.62}$$

Then, we can solve u_1 and u_2 based on the definitions of u_z and u_φ. If we use u_1 and u_2 to replace ones in $\dot{x}_6$ and $\dot{x}_8$, we obtain the following zero dynamics:

$$\dot{x} = Ax + Bz_o + B_r x_r, \tag{4.63}$$

where

$$x = \begin{bmatrix} x_5 \\ x_6 \\ x_7 \\ x_8 \end{bmatrix}, \quad A = \begin{bmatrix} 0 & 1 & 0 & 0 \\ -\frac{k_{f2}}{m_f} & -\frac{b_{f2}}{m_f} & 0 & 0 \\ 0 & 0 & 0 & 1 \\ 0 & 0 & -\frac{k_{r2}}{m_r} & -\frac{b_{r2}}{m_r} \end{bmatrix},$$

$$B = \begin{bmatrix} 0 & 0 & 0 & 0 \\ \frac{k_{f2}}{m_f} & \frac{b_{f2}}{m_f} & 0 & 0 \\ 0 & 0 & 0 & 0 \\ 0 & 0 & \frac{k_{r2}}{m_r} & \frac{b_{r2}}{m_r} \end{bmatrix}, \quad z_o = \begin{bmatrix} z_{o1} \\ \dot{z}_{o1} \\ z_{o2} \\ \dot{z}_{o2} \end{bmatrix},$$

$$B_r = \begin{bmatrix} 0 & 0 \\ -\frac{bM}{m_f(a+b)} & -\frac{I}{m_f(a+b)} \\ 0 & 0 \\ -\frac{bM}{m_r(a+b)} & \frac{I}{m_r(a+b)} \end{bmatrix}, \quad x_r = \begin{bmatrix} \ddot{x}_{1r} \\ \ddot{x}_{3r} \end{bmatrix}.$$

Defining a positive function $V = x^T P x$, with $P > 0$ is a positive matrix, we have

$$\dot{V} = \dot{x}^T P x + x^T P \dot{x} = x^T(A^T P + AP)x + 2x^T PBz_o + 2x^T PB_r x_r.$$

It is easy to verify that the matrix A has eigenvalues with negative real parts. Hence, we have $A^T P + AP = -Q$, where $Q > 0$ is a positive matrix. Noting that

$$2x^T PBz_o \leq \frac{1}{\eta_1} x^T PBB^T Px + \eta_1 z_o^T z_o,$$
$$2x^T PB_r x_r \leq \frac{1}{\eta_2} x^T PB_r B_r^T Px + \eta_2 x_r^T x_r,$$

where η_1, η_2 are tuning positive values, the following inequality is obtained.

$$\begin{aligned}\dot{V} &\leq -x^T Qx + \frac{1}{\eta_1} x^T PBB^T Px + \eta_1 z_o^T z_o \\ &+ \frac{1}{\eta_2} x^T PB_r B_r^T Px + \eta_2 x_r^T x_r \\ &\leq [-\lambda_{\min}(P^{-\frac{1}{2}} QP^{-\frac{1}{2}}) + \frac{1}{\eta_1}\lambda_{\max}(P^{\frac{1}{2}} BB^T P^{\frac{1}{2}}) \\ &+ \frac{1}{\eta_2}\lambda_{\max}(P^{\frac{1}{2}} B_r B_r^T P^{\frac{1}{2}})]V + \eta_1 z_o^T z_o + \eta_2 x_r^T x_r.\end{aligned}$$

Based on properly chosen matrices P, Q and tuning values η_1, η_2, we can find a positive value ε_1, so that

$$\dot{V} \leq -\varepsilon_1 V + \varepsilon_2, \tag{4.64}$$

where $\varepsilon_2 = \eta_1 z_{o\max} + \eta_2 x_{r\max}$ with $z_o^T z_o \leq z_{o\max}$ and $x_r^T x_r \leq x_{r\max}$. (4.64) shows that the Lyapunov function is bounded by

$$V(t) \leq (V(0) - \frac{\varepsilon_2}{\varepsilon_1})e^{-\varepsilon_1 t} + \frac{\varepsilon_2}{\varepsilon_1}, \tag{4.65}$$

which tells us that $|x_k| \leq \sqrt{\frac{q}{\lambda_{\min}(P)}}$, $(k = 5, 6, 7, 8)$ with

$$q = \begin{cases} V(0), & \text{if } V(0) \geq \frac{\varepsilon_2}{\varepsilon_1}, \\ \frac{2\varepsilon_2}{\varepsilon_1} - V(0), & \text{if } V(0) < \frac{\varepsilon_2}{\varepsilon_1}. \end{cases}$$

Performance Constraints

From the analysis above, we know all the signals are bounded within the known ranges, and the bounds of dynamic tire loads can be estimated as

$$|D_f| \leq (k_{f2} + b_{f2})\sqrt{\frac{q}{\lambda_{\min}(P)}} + k_{f2}\,\|z_{o1}\|_\infty + b_{f2}\,\|\dot{z}_{o1}\|_\infty,$$
$$|D_r| \leq (k_{r2} + b_{r2})\sqrt{\frac{q}{\lambda_{\min}(P)}} + k_{r2}\,\|z_{o2}\|_\infty + b_{r2}\,\|\dot{z}_{o2}\|_\infty.$$

Furthermore, the bounds of suspension spaces can be obtained as

$$
\begin{aligned}
\left|\Delta y_f\right| &\le |x_1| + a\,|\sin x_3| + |x_5| \le |x_1| + a\,|x_3| + |x_5| \\
&\le \|x_{1r}\|_\infty + \sqrt{2V_2(0)} + a\,\|x_{3r}\|_\infty \\
&\quad + a\sqrt{2V_4(0)} + \sqrt{\frac{q}{\lambda_{\min}(P)}},
\end{aligned} \tag{4.66}
$$

$$
\begin{aligned}
|\Delta y_r| &\le \|x_{1r}\|_\infty + \sqrt{2V_2(0)} + b\,\|x_{3r}\|_\infty \\
&\quad + b\sqrt{2V_4(0)} + \sqrt{\frac{q}{\lambda_{\min}(P)}}.
\end{aligned} \tag{4.67}
$$

If we adjust the initial values and tuning parameters, then we can always guarantee

$$
\left|D_f\right| \le F_f,\ \ |D_r| \le F_r,\ \ \left|\Delta y_f\right| \le \Delta y_{f\max},\ \ |\Delta y_r| \le \Delta y_{r\max}. \tag{4.68}
$$

Similarly, bounds of $|u_z|$ and $\left|u_\varphi\right|$ can be estimated in the form of

$$
\begin{aligned}
|u_z| &\le M_{\max}(\|\ddot{x}_{1r}\|_\infty + k_1\,|\dot{e}_1| + k_2\,|e_2| + |e_1|) \\
&\quad + \left|F_{df}\right| + |F_{dr}| + \left|F_{sf}\right| + |F_{sr}| \le u_{zbd},
\end{aligned} \tag{4.69}
$$

$$
\begin{aligned}
\left|u_\varphi\right| &\le I_{\max}(\|\ddot{x}_{3r}\|_\infty + k_3\,|\dot{e}_3| + k_4\,|e_4| + |e_3|) \\
&\quad + a(\left|F_{df}\right| + \left|F_{sf}\right|) + b(|F_{dr}| + |F_{sr}|) \le u_{\varphi bd},
\end{aligned} \tag{4.70}
$$

which helps us to get the upper bounds of $|u_1|$ and $|u_2|$ with

$$
|u_1| \le \frac{b u_{zbd} + u_{\varphi bd}}{a+b},\quad |u_2| \le \frac{a u_{zbd} + u_{\varphi bd}}{a+b}. \tag{4.71}
$$

If we adjust the initial values and tuning gains (k_1, k_2, k_3, k_4), then the saturation conditions are satisfied:

$$
|u_1| \le u_{1\max},\quad |u_2| \le u_{2\max}. \tag{4.72}
$$

Remark 4.14 An important problem to be noted is the selection of the initial values and the design parameters. Firstly, according to the analysis above, the initial values should be chosen to satisfy (4.68), which implies that the two hard constraints (suspension spaces and road holding) will be guaranteed. After the initial values are fixed, the gain parameters should be chosen based on (4.69)–(4.70); where the gain parameters should not only ensure the forces u_z and u_φ within their ranges, but also take the control ability into account. Therefore, on the premise that the hard constraints are guaranteed, the gain parameters k_i, ($i = 1, 2, 3, 4$) should be given large values.

Reference Trajectory

In this section, the choice of polynomial in (4.73) can keep the vertical and pitch motions to stabilize in pre-determined time, and also help adjust the corresponding accelerations to high or low levels, which implies that the closed-loop systems can achieve a high performance by making a good trade-off between the pre-determined time and the corresponding accelerations.

$$x_{jr}(t) = \begin{cases} a_{j0} + a_{j1}t + a_{j2}t^2 + a_{j3}t^3 + a_{j4}t^4, & t < T_{jr} \\ 0, & t \geq T_{jr} \end{cases}, \tag{4.73}$$

where $j = 1, 3$, and coefficient vectors $a_{ji}, i = 0, 1, 2, 3, 4,\ j = 1, 3$ are determined such that

$$\begin{aligned} x_{jr}(0) &= a_{j0} = x_j(0), \\ \dot{x}_{jr}(0) &= a_{j1} = x_{j+1}(0), \\ x_{jr}(T_{jr}) &= a_{j0} + a_{j1}T_{jr} + a_{j2}T_{jr}^2 + a_{j3}T_{jr}^3 + a_{j4}T_{jr}^4 = 0, \\ \dot{x}_{jr}(T_{jr}) &= a_{j1} + 2a_{j2}T_{jr} + 3a_{j3}T_{jr}^2 + 4a_{j4}T_{jr}^3 = 0, \\ \ddot{x}_{jr}(T_{jr}) &= 2a_{j2} + 6a_{j3}T_{jr} + 12a_{j4}T_{jr}^2 = 0, \end{aligned} \tag{4.74}$$

which can guarantee that (1) $\dot{e}_1(0) = e_1(0) = 0$; $\dot{e}_3(0) = e_3(0) = 0$; (2) the vector function $x_{jr}(t) \in$ C3. Furthermore, it is easy to see from (4.74) that $x_{jr}(t) = 0$ and $\dot{x}_{jr}(t) = 0$ can be reached in a pre-determined time T_{jr}.

In this section, the proposed adaptive backstepping strategy can realize the multi-objective control for the active suspensions, which means that all the required performances (ride comfort, suspension spaces, road holding and actuator saturation) are considered and improved by our designed controller. Compared with the existing results, most of which just consider partial performances, the proposed controller fully achieves the performances of improved ride comfort, limited suspension spaces, good road holding and allowable actuator inputs. In particular, the choice of polynomial in (4.73) can keep the vertical and pitch motions to stabilize in pre-determined time, and also help adjust the corresponding accelerations to high or low levels, which implies that the closed-loop systems can achieve a high performance by making a good trade-off between the pre-determined time and the corresponding accelerations.

Remark 4.15 In this section, the controllers designed are based on full state feedback. Although the measurements of full states can be realized, this process, sometimes, can be too costly and/or add complexity. It is therefore essential to consider the constrained output feedback control strategy for active suspension systems as a future research target.

4.2.3 Simulation Verification

In this section, we provide an example to illustrate the effectiveness of adaptive backstepping design approach. The half-car model parameters are given as: $M = 1200\,\text{kg}$, $m_f = m_r = 100\,\text{kg}$, $I = 600\,\text{kgm}^2$, $k_{f1} = k_{r1} = 15000\,\text{N/m}$, $k_{nf1} = k_{nr1} = 1000\,\text{N/m}$, $k_{f2} = 200000\,\text{N/m}$, $k_{r2} = 150000\,\text{N/m}$, $b_{f2} = b_e = 1500\,\text{Ns/m}$, $b_{r2} = 2000\,\text{Ns/m}$, $b_c = 1200\,\text{Ns/m}$, $a = 1.2\,\text{m}$, $b = 1.5\,\text{m}$, $V = 20\,\text{m/s}$. Give the initial state values as: $x_1(0) = 3\,\text{cm}$, $x_3(0) = 3\,\text{cm}$, $x_5(0) = 1\,\text{cm}$, $x_7(0) = 1\,\text{cm}$, $\theta_1(0) = 1/1100$, $\theta_2(0) = 1/700$ and the rest ones are assumed as zeros. Besides, to investigate the effect of required settling time T_r on the system response, let the parameters of reference trajectory be $T_r = 0.5\,\text{s}$; $1.0\,\text{s}$; $2.0\,\text{s}$, respectively. The controller parameters are given in Table 4.3.

In this section, the proposed controller was tested by 3 Hz sine wave of 3 cm amplitude. Figure 4.10 shows the time histories of vertical and pitch displacements for both passive systems and active suspensions with adaptive backstepping controllers in the case of the pre-determined settling time $T_r = 0.5\,\text{s}$; $1.0\,\text{s}$; $2.0\,\text{s}$, respectively, and Fig. 4.11 is the corresponding responses of vertical and pitch accelerations. It can be seen that theoretically, we can settle T_r as an arbitrarily small value to make the vertical or pitch displacements vanish fast. However, smaller settle time T_r results in larger vertical and pitch accelerations and larger tracking errors e_1 and e_3, which can be confirmed from Figs. 4.11 and 4.12.

It is well known that the root mean square (RMS) value of the vehicle body acceleration is strictly related to the ride comfort of passengers, and it is often used to quantify the amount of acceleration transmitted to the vehicle body. The RMS value of an n-dimensional vector x is calculated as:

$$x_{RMS} = \frac{\|x\|}{\sqrt{n}} = \sqrt{\frac{1}{n}\sum_{j=1}^{n} x_j^2},\ j = 1, \ldots, n. \tag{4.75}$$

Table 4.4 gives a comparison among the above-mentioned cases using RMS values, and the percentages of the improvement compared to the passive systems are listed in this table, which clearly shows that active suspensions with adaptive controllers improve ride comfort by reducing the effect of road perturbations on the chassis. Furthermore, the fact that larger setting time T_r leads to less RMS value of the acceleration can be confirmed. In the following table, "PSS" and "ASS" stand for passive suspension systems and active suspension systems.

Table 4.3 The controller parameters of active suspensions

Parameter	$r_{1,2}$	$k_{1,2,3,4}$	θ_{1m}	θ_{1M}	θ_{2m}	θ_{2M}
Value	0.001	10	$\frac{1}{1300}$	$\frac{1}{1000}$	$\frac{1}{700}$	$\frac{1}{500}$

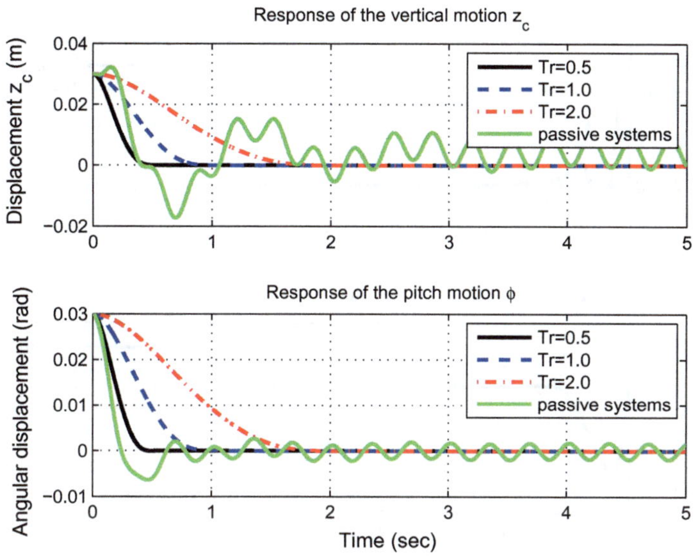

Fig. 4.10 Displacement responses of the vertical and pitch motions with different setting time T_r

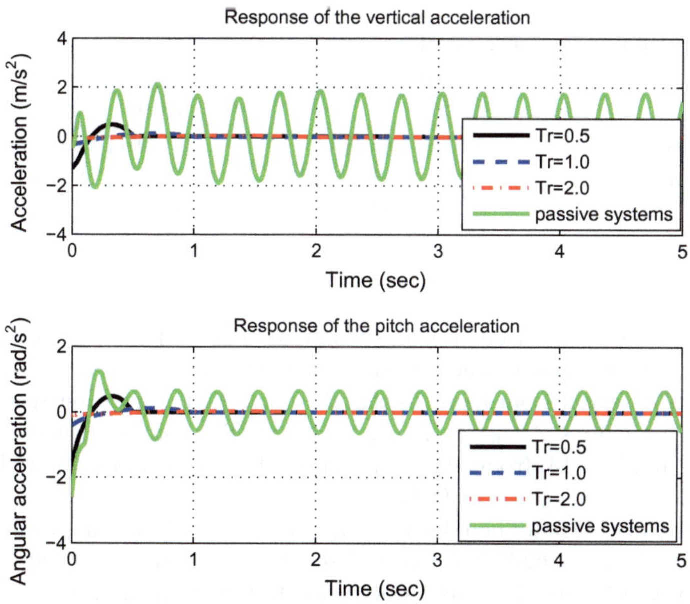

Fig. 4.11 Acceleration responses of the vertical and pitch motions with different setting time T_r

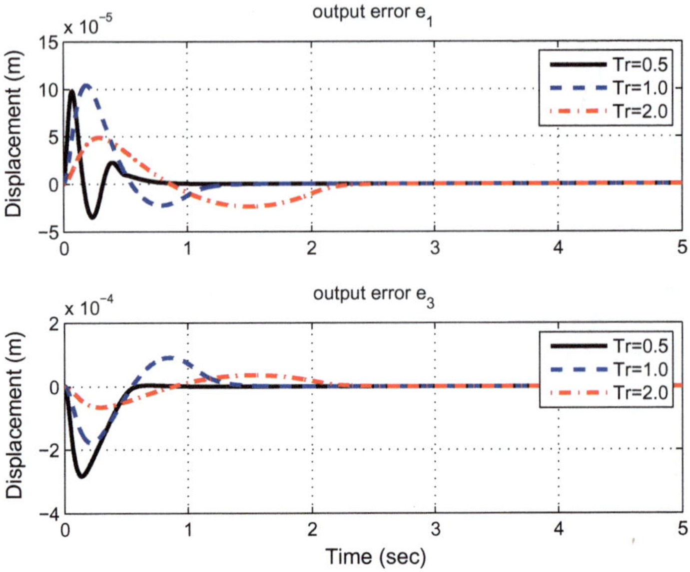

Fig. 4.12 Tracking errors e_1 and e_3 with different setting time T_r

Table 4.4 The RMS values of accelerations

	$\ddot{z}_{crms}$ (m/s^2)	$\ddot{\varphi}_{rms}$ (rad/s^2)
PSS	1.2146	0.5196
ASS ($T_r = 0.5$)	0.1668(↓ 86.27%)	0.1705(↓ 67.19%)
ASS ($T_r = 1$)	0.0587(↓ 95.17%)	0.0594(↓ 88.57%)
ASS ($T_r = 2$)	0.0208(↓ 98.29%)	0.0208(↓ 96.00%)

In the active suspension control, the limitations of the suspension spaces should be taken into account, which means the suspension working space must be preserved. It can been observed from Fig. 4.13 that the controlled suspension spaces are all below the limitations $\Delta y_{f\max} = \Delta y_{r\max} = 0.1\,\text{m}$. The control inputs $u_1(t)$, $u_2(t)$ are plotted in Fig. 4.14, from which we can see that the input forces are below the limitations $u_{1\max} = u_{2\max} = 5000\,\text{N}$.

The static tire loads for both front and rear wheels can be calculated by (4.51) as $F_f = 7513.3\,\text{N}$, $F_r = 6206.7\,\text{N}$. Fig. 4.15 shows the responses of dynamic tire load of the two wheels, and the peaks of dynamic tire load for the two wheels are all within the bounds.

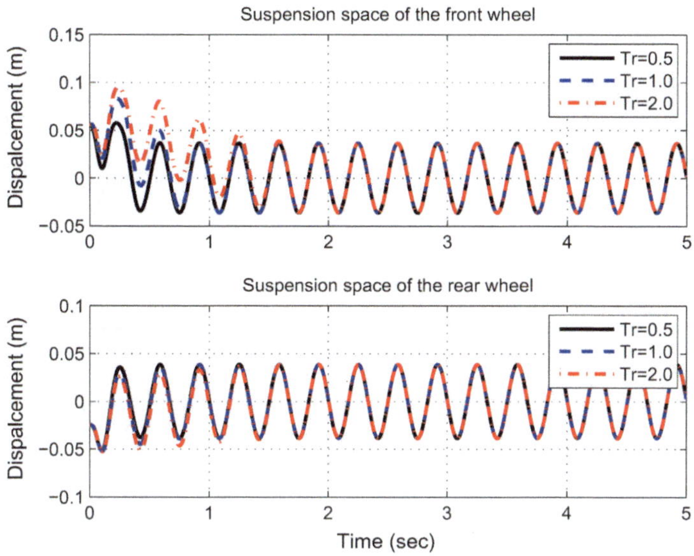

Fig. 4.13 Suspension spaces of the front and rear wheels

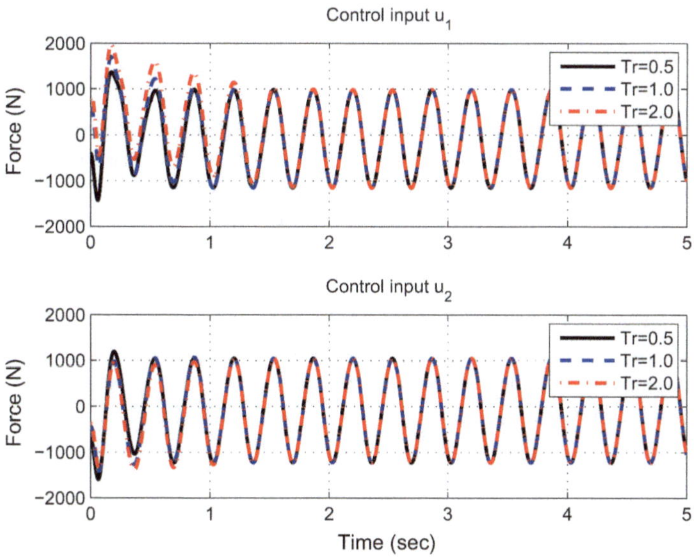

Fig. 4.14 Control inputs u_1 and u_2

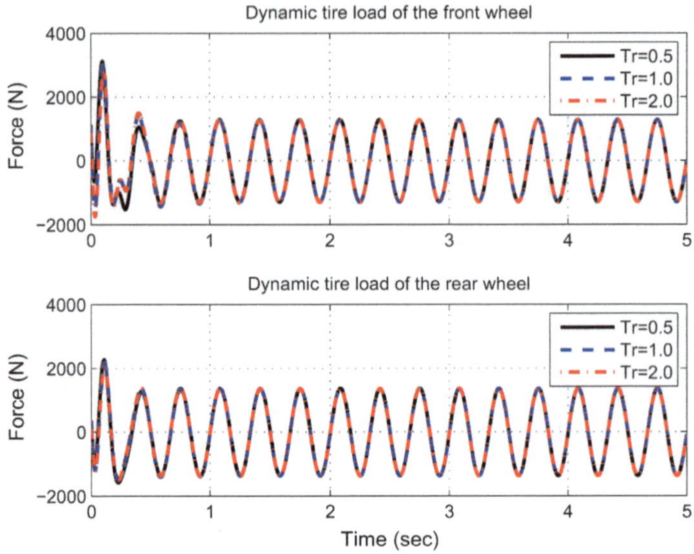

Fig. 4.15 Dynamic tire load of the front and rear wheels

4.2.4 Conclusion

In this section, an adaptive backstepping control strategy has been proposed for vehicle active suspension systems in order to improve ride comfort. On the other hand, the time-domain constraints required in active suspension control have been guaranteed within the whole time domain. By planing a special reference trajectory, the body vertical and pitch displacements can be stabilized in pre-determined time. A half-car model with non-linear spring and piece-wise linear damper has been considered and the effectiveness of the proposed approach has been illustrated by a design example. As future works, it is interesting to consider the integrated control of the several subsystems (such as the active suspension systems, active front steering, anti-locked braking system, and so on) to improve the vehicle dynamic performances.

References

1. M. Mitschke, H. Wallentowitz, *Dynamik der kraftfahrzeuge* (Springer, Berlin, Germany, 2004)
2. S. Bhat, D. Bernstein, Continuous finite-time stabilization of the translational and rotational double integrators
3. Y. Hong, Y. Xu, J. Huang, Finite-time control for robot manipulators. Syst. Control Lett. **46**(4), 243–253 (2002)
4. S. Bhat, D. Bernstein, Finite time stability of homogeneous systems. Am. Control Conf. **1**(6), 2513–2514 (1997)

5. L. Rosier, Homogeneous Lyapunov function for homogeneous continuous vector field. Syst. Control Lett. **19**(6), 467–473 (1992)
6. V.T. Haimo, Finite time controllers. SIAM J. Control Optim. **24**(4), 760–770 (1986)
7. S. Bhat, D. Bernstein, Geometric homogeneity with applications to finite-time stability. Math. Control Signals Syst. **17**(2), 101–127 (2005)
8. S. Bhat, D. Bernstein, Finite-time stability of continuous autonomous systems. SIAM J. Control Optim. **28**(3), 751–766 (2000)
9. M. Canale, M. Milanese, C. Novara, Semi-active suspension control using "Fast" model-predictive techniques. IEEE Trans. Control Syst. Technol. **14**(6), 1034–1046 (2006)
10. Z. Wang, Y. Liu, X. Liu, H_∞ filtering for uncertain stochastic time-delay systems with sector-bounded nonlinearities. Automatica **44**(5), 1268–1277 (2008)
11. H. Karimi, H. Gao, New delay-dependent exponential H_∞ synchronization for uncertain neural networks with mixed time delays. IEEE Trans. Syst. Man Cybern. Part B Cybern. **40**(1), 173–185 (2010)
12. P. Shi, Filtering on sampled-data systems with parametric uncertainty. IEEE Trans. Automat. Control **43**(7), 1022–1027 (1998)
13. B. Yao, F. Bu, J. Reedy, G. Chiu, Adaptive robust motion control of single-rod hydraulic actuators: theory and experiments. IEEE Trans. Mechatron. **5**(1), 79–91 (2000)
14. B. Yao, M. Tomizuka, Adaptive robust control of MIMO nonlinear systems in semi-strict feedback forms. Automatica **37**, 1305–1321 (2001)
15. J. Slotine, W. Li, *Applied Nonlinear Control* (Prentice-Hall, Englewood Cliffs, NJ, 1991)

Chapter 5
Actuator Saturation Control for Active Suspension Systems

Actuator saturation is very common in an actual control system, and has proved a source of performance degradation and even instability. In practice, it is not always possible to ensure that all signals are small, particularly for high-performance application, and meantime, actuators which deliver the control signal in physical applications are always subject to the magnitude or rate limits. This chapter provides some control strategies about saturation problems. In Sect. 5.1, a saturated adaptive robust control (ARC) strategy is proposed to dealing with the possible actuator saturation. In Sect. 5.2, we develop a controller solution for active suspension systems considering parameter uncertainties, performance constraints, and actuator saturation problems, which is aimed at vibration isolation.

5.1 Saturated Adaptive Robust Control for Active Suspension Systems

This section investigates the problem of vibration control in vehicle active suspension systems, whose aim is to stabilize the attitude of vehicle and improve ride comfort. In response to uncertainties in systems and the possible actuator saturation, an ARC strategy is proposed. Specifically, an anti-windup block is added to adjust the control strategy in a manner conductive to stability and performance preservation in presence of saturation. Furthermore, the proposed saturated ARC approach is applied to the half-car active suspension systems, where nonlinear springs and piece-wise linear dampers are adopted. Finally, the typical bump road inputs are considered as the road disturbances in order to illustrate the effectiveness of the proposed control law.

In this section, the nonlinear half-car model is the same model in Sect. 4.2. This model has been used extensively in the literature and captures many important characteristics of vertical and pitch motions. Here, the effect of lateral motion is neglected. The definitions of state variables and the state-space expression of the nonlinear half-car model is derived as (4.49) in Sect. 4.2.

W. Sun et al., *Advanced Control for Vehicle Active Suspension Systems*, Studies in Systems, Decision and Control 204,
https://doi.org/10.1007/978-3-030-15785-2_5

5.1.1 Problem Statement

In this section, the main task is to design a controller which can succeed in stabilizing the vertical and pitch motion of the car body and isolating the force transmitted to the passengers as well. Additionally, actuators of active suspensions are likely subject to amplitude saturation with the saturation bound $u_{\max}$, that is $|u_i(t)| \leq u_{\max}$, $i = 1, 2$. Consequently, this study tries to deal with the following control problem.

Problem 5.1 For the active suspension systems, synthesize adaptive control inputs u_i $(i = 1, 2)$ to stabilize the heave and pitch motions of closed-loop systems in the presence of parametric uncertainties and uncertain nonlinearities, and meanwhile the stability of the closed-loop systems can be maintained and less performance degradation can be obtained when the actuator saturation occurs.

5.1.2 Saturated ARC Controller Synthesis

First, the control function u_z for the heave motion is designed to keep the tracking error $e_1 = x_1 - x_{1r}$ to converge to zero or bounded, where x_{1r} represents the reference trajectory. The governing equations for the heave motion are:

$$\begin{aligned} \dot{x}_1 &= x_2, \\ \dot{x}_2 &= \theta_1 (F(x,t) + \text{sat}_{u_{z\max}}(u_z)) + d_1(x,t), \end{aligned} \tag{5.1}$$

where $F(x,t) = -F_{df} - F_{dr} - F_{sf} - F_{sr}$ and $\theta_1 = \frac{1}{M} \in \left[\theta_{1\min}\ \theta_{1\max}\right]$. In this study, the scalar saturation function of level $a \in \mathbb{R} > 0$ is defined as $\text{sat}_a(s) :=\text{sign}(s)\min\{|s|, a\}$, and the function $\text{sat}_{a,b}(s)$ is defined as

$$\text{sat}_{a,b}(s) := \begin{cases} a, \ s < a, \\ s, \ a \leq s \leq b, \\ b, \ s > b. \end{cases}$$

The basic idea underlying anti-windup designs with saturating actuators is to introduce control modifications in order to recover, as much as possible, the performance induced by a previous design carried out on the basis of the unsaturated system. Here, the design process of unsaturated adaptive robust controller is given firstly in the following lemma.

Lemma 5.2 *When the saturation does not occur, for the system in (5.1), with the controller law (5.2)–(5.4) and the projection type adaptive law (5.5), the following results hold:*

A. In general (i.e., the system is subjected to parametric uncertainties, unmodelled uncertainties and external disturbances), all signals in system (5.1) are bounded;

B. If after a finite time, the system is subjected to parametric uncertainties only (i.e., all the disturbances vanish after a finite time), the tracking error will converge to zero in a finite time.

$$u_z = u_{za} + u_{zs}, \tag{5.2}$$

$$u_{za} = \frac{1}{\hat{\theta}_1}(\dot{x}_2^d - e_1 - k_2 e_2) - F(x,t), \tag{5.3}$$

$$u_{zs} = -\frac{h_0^2(x,t)}{4\theta_{1\,\min}\varepsilon_0} e_2, \tag{5.4}$$

$$\dot{\hat{\theta}}_1 = Proj_{\hat{\theta}_1}(r_0\tau) = \begin{cases} 0, \text{ if } \hat{\theta}_1 = \theta_{1\,\max}, r_0\tau > 0, \\ 0, \text{ if } \hat{\theta}_1 = \theta_{1\,\min}, r_0\tau < 0, \\ r_0\tau, \text{ otherwise,} \end{cases} \tag{5.5}$$

where

$$e_1 = x_1 - x_{1r},\ e_2 = x_2 - x_2^d,\ x_2^d = \dot{x}_{1r} - k_1 e_1,$$

and $r_0 > 0$ is a tunable gain, $\tau = (F(x,t) + u_z)e_2$, $\varepsilon_0 > 0$ is a constant value and $h_0(x,t)$ is a bounded function which is chosen based on

$$\tilde{\theta}_1(F(x,t) + u_z) + d_1 \le \left|\tilde{\theta}_1\right| |F(x,t) + u_z| + \|d_1\|_\infty = h_0(x,t).$$

Proof First, the error systems can be given as:

$$\dot{e}_1 = e_2 - k_1 e_1. \tag{5.6}$$

$$\dot{e}_2 = \theta_1(F(x,t) + u_z) + d_1(x,t) - \dot{x}_2^d, \tag{5.7}$$

where $x_2^d = \dot{x}_{1r} - k_1 e_1$. Selecting a Lyapunov candidate function as $V = \frac{1}{2}e_1^2 + \frac{1}{2}e_2^2$, we have

$$\dot{V} = e_1(e_2 - k_1 e_1) + e_2(\theta_1(F(x,t) + u_z) + d_1(x,t) - \dot{x}_2^d).$$

Substituting the controller law (5.34)–(5.4) into the equation above results in

$$\begin{aligned} \dot{V} &= -k_1 e_1^2 - k_2 e_2^2 + e_2(-\frac{\hat{\theta}_1 h_0^2 e_2}{4\theta_{1\,\min}\varepsilon_0} - \tilde{\theta}_1(F(x,t) + u_z) + d_1) \\ &\le -k_1 e_1^2 - k_2 e_2^2 + \varepsilon_0. \end{aligned}$$

After defining $\lambda_0 = \min(k_1, k_2)$, we have

$$\dot{V} \le -2\lambda_0 V + \varepsilon_0,$$

which can further result in

$$V(t) \leq \frac{\varepsilon_0}{2\lambda_0} + (V(0) - \frac{\varepsilon_0}{2\lambda_0})e^{-2\lambda_0 t}. \tag{5.8}$$

Inequality (5.8) shows that e_1, e_2 are bounded as $t \to \infty$.

If the disturbances vanish after a finite time ($d_1(x,t) = 0$), differentiating V results in

$$\dot{V} \leq -k_1 e_1^2 - k_2 e_2^2 - \tilde{\theta}_1(F(x,t) + u_z)e_2. \tag{5.9}$$

Noticing the property of $\tilde{\theta}_1(r_0^{-1}\dot{\hat{\theta}}_1 - \tau) \leq 0$, $\forall \tau$, and selecting the positive semi-definite function

$$V_a = V + \frac{1}{2r_0}\tilde{\theta}_1^2,$$

we have

$$\dot{V}_a \leq -k_1 e_1^2 - k_2 e_2^2 \leq 0,$$

which implies that e_1, e_2 converge to zero as $t \to \infty$, by using barbalat's Lemma.

In this section, when the saturation is considered, an anti-windup compensation block shown in Fig. 5.1 is used to modify the closed-loop's behavior such that it is more resilient to saturation. The anti-windup compensator in this paper are composed of two filters. The function of the first-order linear filter in (5.10) is to modify the error e_2,

$$\begin{aligned} e_2 &= x_2 - x_2^d + \eta_1, \\ \dot{\eta}_1 &= -k_{01}\eta_1 + \theta_{10}\Delta u_z, \end{aligned} \tag{5.10}$$

and the nonlinear filter with the form of (5.11) is added to reduce the magnitude of the control input u_{za}, when the saturation is encountered.

$$\begin{aligned} \dot{\xi}_1 &= f(\xi_1, \Delta u_z), \\ v_1 &= g(\xi_1, \Delta u_z). \end{aligned} \tag{5.11}$$

Functions $f(\xi_1, \Delta u_z)$ and $g(\xi_1, \Delta u_z)$ are then chosen in such a way that when saturation occurs, v_1 rises rapidly to 1 to force the controller output into the linear region of the saturation in [1]. When the controller output is small enough, v_1 goes slowly back to zero, thus recovering the unsaturated closed-loop dynamics while preserving global boundness. This is achieved by the following selections:

$$\begin{aligned} f(\xi_1, \Delta u_z) &= \mathrm{sat}_{-c^-,c^+}(k_l(\mathrm{sat}_{0,m}(k_l\,|\Delta u_z|) - \xi_1)), \\ g(\xi_1, \Delta u_z) &= \mathrm{sat}_{0,1}(\xi_1), \end{aligned} \tag{5.12}$$

where c^-, c^+, m, and k_l are the positive scalar parameters.

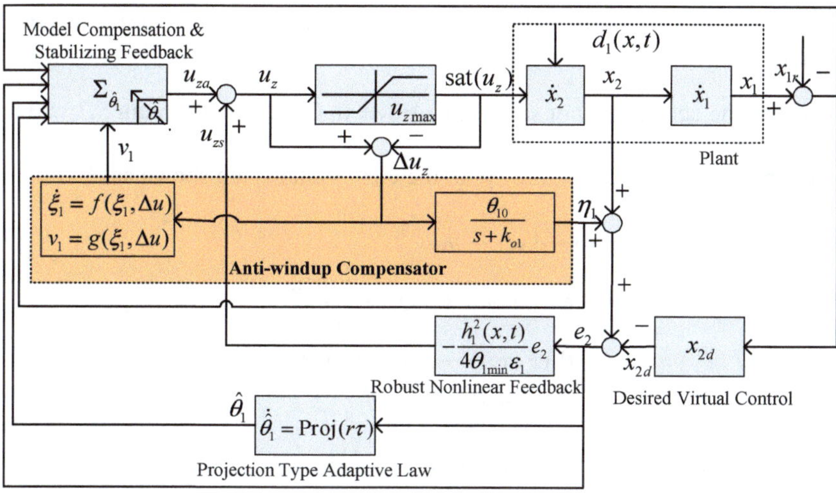

Fig. 5.1 Structure diagram of saturated ARC controller design based on anti-windup strategy

Theorem 5.3 *For the system in (5.1), with the modified ARC law (5.13)–(5.16) and the projection type adaptive law (5.17), the following results hold:*

Case 1. *When the saturation does not occur,* $\Delta u_z = 0$*, the closed-loop systems function the same dynamics as ones in Lemma 1, and all the solutions in Lemma 1 are still hold automatically here.*

Case 2. *When the saturation occurs,* $\Delta u_z \neq 0$*, the tracking error* e_1 *is still bounded, and the tracking performance can be maintained and less performance degradation occurs compared with the closed-loop systems without anti-windup block.*

$$u_z = u_{za} + u_{zs}, \tag{5.13}$$

$$u_{za} = \frac{1}{\hat{\theta}_1}(\dot{x}_2^d + k_{o1}\eta_1 - e_1 - (k_2 - v_1 k_{a1})e_2) - F(x,t),$$

$$u_{zs} = -\frac{h_1^2(x,t)}{4\theta_{1\min}\varepsilon_1}e_2,$$

$$\dot{\eta}_1 = -k_{01}\eta_1 + \theta_{10}\Delta u_z, \tag{5.14}$$

$$\dot{\xi}_1 = \mathrm{sat}_{-c^-,c^+}(k_l(\mathrm{sat}_{0,m}(k_l\,|\Delta u_z|) - \xi_1)), \tag{5.15}$$

$$v_1 = \mathrm{sat}_{0,1}(\xi_1), \tag{5.16}$$

$$\dot{\hat{\theta}}_1 = Proj_{\hat{\theta}_1}(r_1(F(x,t) + u_z)e_2), \tag{5.17}$$

where $k_2 > k_{a1} > 0$, $h_1(x,t)$ can be chosen as

$$\begin{aligned}&\tilde{\theta}_1(F(x,t) + u_z) + \bar{d}_1(x,t)\\ &\leq |\theta_{1\max} - \theta_{1\min}|\,|F(x,t) + u_z| + \left\|\bar{d}_1\right\|_\infty = h_1(x,t),\end{aligned}$$

with $\bar{d}_1(x,t) = d_1(x,t) + (\theta_{10} - \theta_1)\Delta u_z$, and θ_{10} is the initial value of the uncertain parameter θ_1.

Proof When the actuator is not saturated, $\Delta u_z = 0$, Case 1 of Theorem 5.3 will be simplified as Lemma 5.2.

When the saturation occurs, $\Delta u_z \neq 0$, the error systems are shown as:

$$\begin{aligned}\dot{e}_1 &= e_2 - k_1 e_1 - \eta_1.\\ \dot{e}_2 &= -e_1 - (k_2 - k_{a1})e_2 - \frac{\hat{\theta}_1 h_1^2(x,t)}{4\theta_{1\min}\varepsilon_1}e_2\\ &\quad -\tilde{\theta}_1(F(x,t) + u_z) + \bar{d}_1(x,t),\\ \dot{\eta}_1 &= -k_{01}\eta_1 + \theta_{10}\Delta u_z.\end{aligned}$$

Selecting a Lyapunov candidate function as

$$V_2(e,\eta_1) = \frac{1}{2}e_1^2 + \frac{1}{2}e_2^2 + \frac{1}{2}\eta_1^2,$$

obviously, for any $\gamma > 0$, the set $Br = \{e, \eta_1 : V_2(e,\eta_1) \leq \gamma\}$ is a compact set. It is worthwhile noting that on the set B_r, $\|\Delta u_z\|$ has the maximum. Note that

$$-e_1\eta \leq e_1^2 + \frac{1}{4}\eta^2,\ \theta_{10}\eta_1\Delta u_z \leq \theta_{10}\eta_1^2 + \mu_1,$$

where $\mu_1 = \frac{1}{4}\theta_{10}\Delta u_z^2$. Calculating the derivative, we have

$$\begin{aligned}\dot{V}_2 &\leq -k_1 e_1^2 - (k_2 - k_a)e_2^2 - k_{01}\eta_1^2 - e_1\eta\\ &\quad +\theta_{10}\eta_1\Delta u_z + (-\frac{h_1^2(x,t)}{4\varepsilon_1}e_2^2 + h_1(x,t)\,|e_2|)\\ &\leq -(k_1 - 1)e_1^2 - (k_2 - k_a)e_2^2 - (k_{01} - \theta_{10} - \frac{1}{4})\eta^2 + \sigma_1.\end{aligned}$$

where $\sigma_1 = \varepsilon_1 + \mu_1$. Choosing $\lambda_1 = \min\{k_1 - 1, k_2 - k_a, k_0 - \theta_{10} - \frac{1}{4}\}$, we can obtain that

$$\dot{V} \leq -2\lambda_1 V + \sigma_1,$$

which further results in

$$V(t) \leq \frac{\sigma_1}{2\lambda_1} + (V(0) - \frac{\sigma_1}{2\lambda_1})e^{-2\lambda_1 t}. \tag{5.18}$$

Inequality (5.18) shows that e_1, e_2, η_1 are bounded as $t \to \infty$. The proof has been finished.

Following a similar procedure, the resultant control function u_φ for the pitch motion of the vehicle body can be designed based on the following equations.

$$\begin{aligned}\dot{x}_3 &= x_4,\\ \dot{x}_4 &= \theta_2(G(x,t) + \mathrm{sat}_{u_{\varphi\max}}(u_\varphi)) + d_2(x,t),\end{aligned}$$

where $G(x,t) = -a(F_{df} + F_{sf}) + b(F_{dr} + F_{sr})$ and $\theta_2 = \frac{1}{I} \in \left[\theta_{2\min}\ \theta_{2\max}\right]$. After obtaining u_z and u_φ, we can calculate the real inputs u_1 and u_2 as

$$u_1 = \frac{bu_z + u_\varphi}{a+b},\ u_2 = \frac{au_z - u_\varphi}{a+b}. \tag{5.19}$$

Remark 5.4 The ARC design yields to a 4th order error dynamic, while the original system is a 8th order system. So the zero dynamics consists of four states. To find it, we set $e_1 = e_3 = 0$, which implies $e_2 = e_4 = 0$. Because actuator saturation is caused by large errors e_i $(i = 1...4)$, the saturation should not occur when $e_i = 0$, that is to say $\eta_1 = \eta_2 = 0$. Hence, we obtain:

$$\begin{aligned}u_z &= M(\ddot{x}_{1r} - d_1) - F(x,t),\\ u_\varphi &= I(\ddot{x}_{3r} - d_2) - G(x,t).\end{aligned} \tag{5.20}$$

Then, we can solve u_1 and u_2 based on the definitions of u_z and u_φ.

$$\begin{aligned}u_1 &= \frac{bM}{a+b}(\ddot{x}_{1r} - d_1) + \frac{I}{a+b}(\ddot{x}_{3r} - d_2) + F_{df} + F_{sf},\\ u_2 &= \frac{bM}{a+b}(\ddot{x}_{1r} - d_1) - \frac{I}{a+b}(\ddot{x}_{3r} - d_2) + F_{dr} + F_{sr}.\end{aligned} \tag{5.21}$$

If we use u_1 and u_2 in (5.21) to replace ones in $\dot{x}_6$ and $\dot{x}_8$, we obtain the following zero dynamics:

$$\dot{\chi} = A\chi + Bz_o + B_r x_r + B_d d, \tag{5.22}$$

where

$$x_r = \begin{bmatrix}\ddot{x}_{1r}\\ \ddot{x}_{3r}\\ d_1\\ d_2\end{bmatrix},$$

$$B_r = \begin{bmatrix} 0 & 0 & 0 & 0\\ -\frac{bM}{m_f(a+b)} & -\frac{I}{m_f(a+b)} & \frac{bM}{m_f(a+b)} & \frac{I}{m_f(a+b)}\\ 0 & 0 & 0 & 0\\ -\frac{bM}{m_r(a+b)} & \frac{I}{m_r(a+b)} & \frac{bM}{m_r(a+b)} & -\frac{I}{m_r(a+b)} \end{bmatrix}.$$

Hence, we easily obtain that the zero dynamics are stable for the reason that the matrix A is Hurwitz.

Remark 5.5 As it is well known, the suspension space is used by the actuators in order to compensate for the road-induced vehicle body vibrations in active suspensions. Thus, if absolute zero reference is assigned to the heave motion of the vehicle body, it will diminish the suspension working space or cause it to reach to the working limits in [2]. Therefore, in order to preserve the suspension working limits, it is assumed that the raw reference value for the heave motion of the vehicle body is equal to the effective value of the unsprung mass displacements under the center of gravity of the vehicle body, that is:

$$x_{1r} = \frac{az_2 + bz_1}{a + b}. \tag{5.23}$$

5.1.3 Simulation Verification

To illustrate the effectiveness of the proposed controller, a half-car model parameters are listed in Table 5.1.

Give the initial state values as: $x_1(0) = 10\text{cm}$, $x_3(0) = 5\text{cm}$, $\theta_1(0) = 1/1250$, $\theta_2(0) = 1/550$ and the rest ones are assumed as zeros. Here, in this simulation, it is assumed that the disturbance nonlinear items $d_1(x, t)$ and $d_2(x, t)$ are 4 Hz sine signals to verify the effectiveness of the proposed ARC controller, that is to say: $d_1(x, t) = d_2(x, t) = \sin(8\pi t)$. The controller parameters are given in Table 5.2.

In order to evaluate the suspension characteristics with respect to ride comfort and actuator saturation, the variability of the road profiles is taken into account. In the context of vehicle suspension performance, road disturbances can be generally assumed as discrete events of relatively short duration and high intensity, caused by, for example, a pronounced bump or pothole on an otherwise smooth road. In the following, a kind of road profile is used to validate the performance of the presented

Table 5.1 The model parameters of half-car active suspensions

Parameter	Value	Parameter	Value
M	1200 kg	b_{f2}	1500 Ns/m
$m_f = m_r$	100 kg	b_{r2}	2000 Ns/m
I	600 kgm^2	b_e	1500 Ns/m
$k_{f1} = k_{r1}$	15000 N/m	b_c	1200 Ns/m
$k_{nf1} = k_{nr1}$	1000 N/m	a	1.2 m
k_{f2}	200000 N/m	b	1.5 m
k_{r2}	150000 N/m	V	20 m/s

Table 5.2 The controller parameters of active suspensions

Parameter	Value	Parameter	Value
$r_1 = r_2$	0.001	$k_1 = k_3$	10
$k_2 = k_4$	40	$k_{o1} = k_{o2}$	10
k_a	39	k_l	10
$u_{1\max} = u_{2\max}$	1500 N	m	2
$h_1 = h_2$	100	c^+	100
c^-	0.1	$\theta_{1\max}$	1/1000
$\theta_{1\min}$	1/1300	$\theta_{2\max}$	1/500
$\theta_{2\min}$	1/700		

control approach. Now consider the case of an isolated bump in an otherwise smooth road surface. The corresponding ground displacement is given by

$$z_{o1} = \begin{cases} \frac{h_b[1-\cos(8\pi t)]}{2}, & 1 \le t \le 1.25, \\ 0, & \text{otherwise}, \end{cases} \tag{5.24}$$

where h_b is the height of the bump road input. It is assumed that the bump road input has the magnitude for $h_b = 2$cm, and road conditions for the front and rear wheels are the same but with a time delay of $(a+b)/V$, where V is the velocity of the vehicle. For subsequent comparison, three kinds of closed-loop responses are plotted:

S1: Saturated closed-loop systems with ARC controller and anti-windup block;
S2: Saturated closed-loop systems with ARC controller, but without anti-windup block;
S3: Unconstrained closed-loop systems with ARC controller.

Figure 5.2 shows the time histories of vertical displacements for the above-mentioned three systems, and it can be seen that our proposed controller can stabilize the vertical and pitch motion better in spite of the presence of input saturation. On the contrary, if we don't use the anti-windup compensator (case S2), there is more performance degradation than the systems with anti-windup block, which implies that our proposed approach works well. In these figures, we can see the vertical and pitch displacements are uniformly ultimately bounded, and it is worthy to mention that if the disturbance nonlinear items $d_1(x,t)$ and $d_2(x,t)$ vanish after a finite time, then all the signals will converge to zero.

Figures 5.3 and 5.4 are plotted to show the tracking errors e_1, e_3, and the corresponding control input functions u_z and u_φ, from which we can see that our proposed approach can obtain the less tracking errors compared with the saturated system without the anti-windup compensator.

In active suspension control, it is widely accepted that ride comfort is closely related to the body acceleration. Here, the time-domain responses of body vertical and pitch acceleration for the active suspension system are shown in Fig. 5.5, where

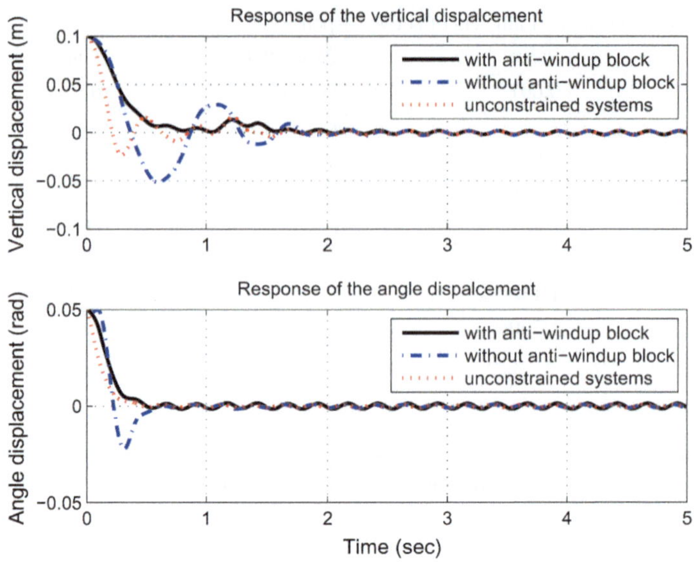

Fig. 5.2 Displacement responses of the vertical and pitch motions

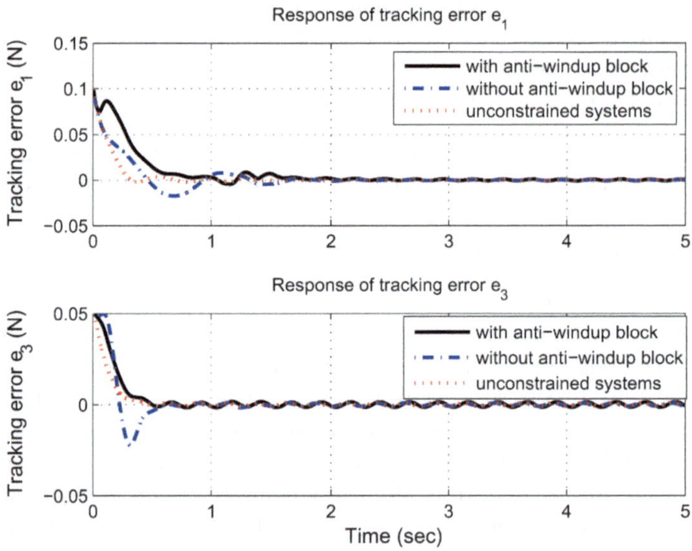

Fig. 5.3 Tracking errors e_1 and e_3

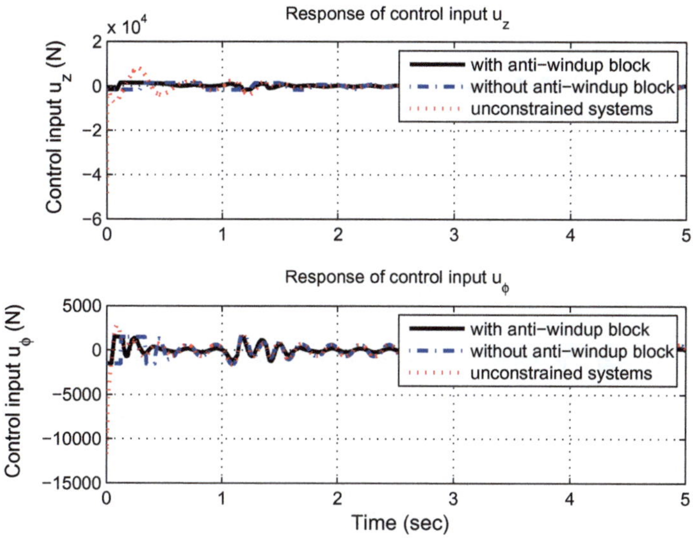

Fig. 5.4 Control function u_z and u_φ

the black solid lines and the blue dashed lines are the responses of acceleration with anti-windup ARC controller and ARC controller without anti-windup compensator, respectively. The red dotted lines represent the unconstrained systems with ARC controller. It is observed from these figures that our proposed approach improves suspension performances in terms of peak response values compared with the other two systems, even though the saturation is encountered. The results confirm the efficiency of our designed controllers.

Besides, the limitations of the suspension spaces should be taken into account, which means the suspension working space must be preserved. It can been observed from Fig. 5.6 that the controlled suspension spaces both fall into the acceptable ranges, whose maximums are 0.16 m and 0.05 m, respectively.

Figure 5.7 is plotted here to show the stability of the zero dynamic systems, from which we can see the states of the zero dynamic systems are stable, and furthermore, our proposed approach can obtain the least peak response values of the unsprung mass displacement among the three comparisons.

In this section, a saturated ARC strategy has been proposed for a nonlinear active suspension system with saturated inputs. After designing a nominal ARC controller for the unconstrained nonlinear systems with parameter uncertainties and external disturbances, an anti-windup compensator has been added to adjust the control strategy and then to match stability and performance preservation in presence of saturation. Finally, the saturated ARC approach has been applied to the half-car active suspensions to illustrate the effectiveness of the proposed control law.

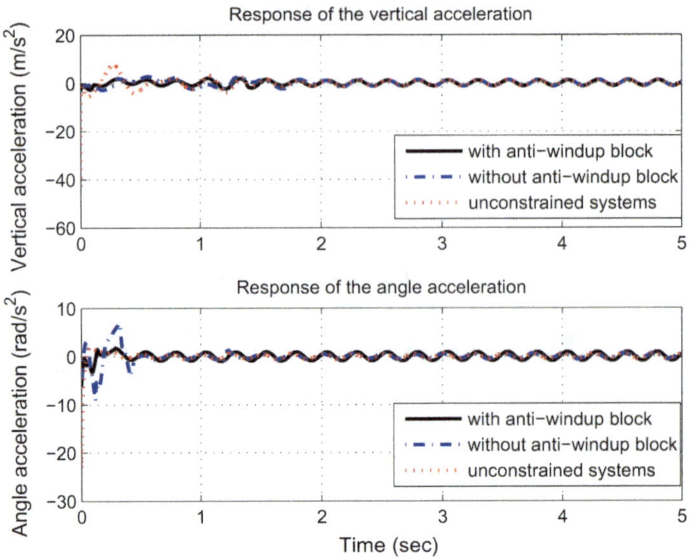

Fig. 5.5 Acceleration responses of the vertical and pitch motions

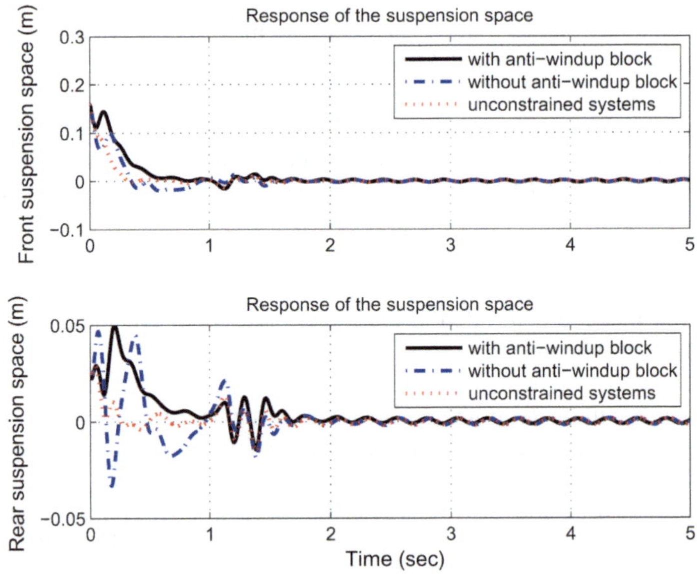

Fig. 5.6 Time responses of the front and rear suspension spaces

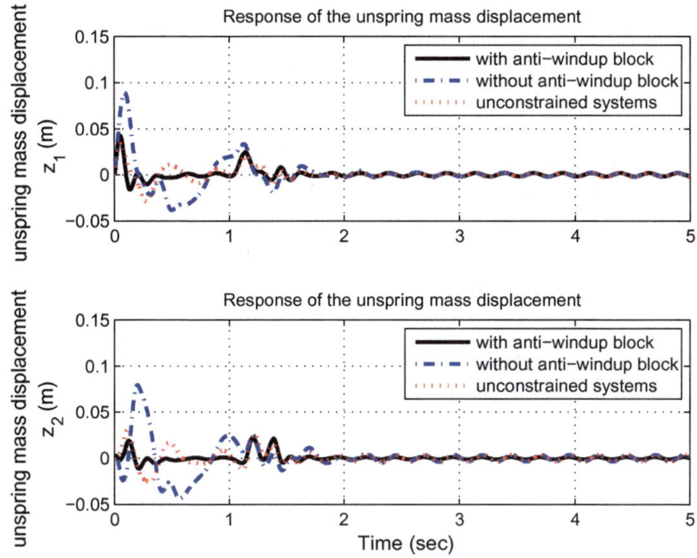

Fig. 5.7 Time responses of unsprung masses displacement z_1 and z_2

5.2 Vibration Isolation for Active Suspensions with Performance Constraints and Actuator Saturation

This section investigates the problem of vibration isolation for vehicle active suspension systems, where parameter uncertainties, external disturbances, actuator saturation and performance constraints are considered in an unified framework. A constrained adaptive robust control technology is proposed to not only stabilize the attitude of vehicle in the context of parameter uncertainties and external disturbances, but also cover the problems of actuator saturation and performance constraints. In addition to spring nonlinearity, the piece-wise linear behavior of the damper, which has different damping rates for compression and extension movements, is taken into consideration to form the basis of accurate control. Furthermore, the performance analysis of the closed-loop systems is given, by means of rigorous mathematical derivations. Extensive comparative experimental results are obtained to illustrate the effectiveness of the proposed control law.

5.2.1 Problem Formulation

In this section, the nonlinear quarter-car model is considered, which is the same as the one shown in Fig. 4.1 in Sect. 4.1. The dynamic equations of the sprung and unsprung masses are given by:

$$
\begin{aligned}
m_s\ddot{z}_s + F_d + F_s &= u + F_l, \\
m_u\ddot{z}_u - F_d - F_s + F_t + F_b &= -u,
\end{aligned}
\tag{5.25}
$$

where F_l denotes the friction force of suspension components, and the forces produced by the non-linear stiffening spring, the piece-wise linear damper and the tire obey the same expressions as (4.2).

Defining the state variables $x_1 = z_s$, $x_2 = \dot{z}_s$, $x_3 = z_u$, $x_4 = \dot{z}_u$ and considering the situation of actuator saturation, the dynamic equations in (5.25) can be rewritten in the following state-space form:

$$
\begin{aligned}
\dot{x}_1 &= x_2, \\
m_s\dot{x}_2 &= \psi(x,t) + \text{sat}(u) + F_l, \\
\dot{x}_3 &= x_4, \\
m_u\dot{x}_4 &= -\psi(x,t) - F_t - F_b - \text{sat}(u), \\
y &= x_1,
\end{aligned}
\tag{5.26}
$$

where $\psi(x,t) = -F_d - F_s$, and

$$\text{sat}(u) = \text{sign}(u) \cdot \min\{|u|, u_{\max}\}$$

means the saturation function of input u.

As we know, the body mass m_s usually changes with the vehicle load, which results in the model containing the uncertain parameter. In this section, we assume that

$$m_s \in \Omega_m = \{m_s : m_{s\min} \leq m_s \leq m_{s\max}\},$$

where $m_{s\min}$ and $m_{s\max}$ are the lower and upper bounds. In addition, we assume that

$$|F_l| \leq d,$$

where d is a constant value, and

$$\phi_{\min}(x,t) \leq \psi(x,t) \leq \phi_{\max}(x,t),$$

$$|\phi_{\min}(x,t)| \leq h, \quad |\phi_{\max}(x,t)| \leq h,$$

with is a constant positive value h.

For active suspension systems, the performance requirements to be considered in the controller design include the following aspects.

1. Ride comfort: in active suspension design, the main task is to design a controller which can succeed in stabilizing the vertical motion of the car body and isolating the force transmitted to the passengers as well.

2. Good road holding: in order to make sure the car safety, we should ensure the firm uninterrupted contact of wheels to road, and the dynamic tire load should be small, that is,
$$|F_t + F_b| < (m_{s\min} + m_u)g. \tag{5.27}$$
3. Suspension space limit: because of mechanical structure, the suspension space should not exceed the allowable maximums, which can be described as
$$|z_s - z_u| \leq z_{\max}, \tag{5.28}$$
where $z_{\max}$ is the maximum suspension deflection.
4. Actuator saturation: all actuators of physical devices are subject to amplitude saturation. Although, in some applications, it may be possible to ignore this fact, the reliable operation and acceptable performance of most control systems must be assessed in light of actuator saturation.

Based on the above statements, this study tries to deal with the following constrained adaptive robust control problem:

Problem 5.6 For the active suspension systems, synthesize a control input u to stabilize the vertical motion of closed-loop systems in the presence of parametric uncertainties and external disturbances. Meanwhile, the essential performance constraints including the ride safety condition in (5.27) and suspension space limit in (5.28) are guaranteed, and the high performances under actuator saturation are maintained.

5.2.2 *Control Law Synthesis*

Controller Design

In order to stabilize the vertical motion y, denote $z_1 = y - y_d$ as the tracking error, where y_d is a reference trajectory which is of convergence to zero as t goes to the infinite. Clearly, the vertical motion y will be stabilized, as long as z_1 converges to zero or be bounded. By means of backstepping technology, starting with the equation of tracking error, we have
$$\dot{z}_1 = x_2 - \dot{y}_d. \tag{5.29}$$

Design a desired virtual control α_1, and let z_2 be an error variable representing the difference between the actual and virtual control of (5.29), i.e., $z_2 = x_2 - \alpha_1$. Thus we can rewrite (5.29) as
$$\dot{z}_1 = z_2 + \alpha_1 - \dot{y}_d. \tag{5.30}$$

To response to the actuator saturation, following the approach proposed in [3], the desired virtual control α_1 is proposed as

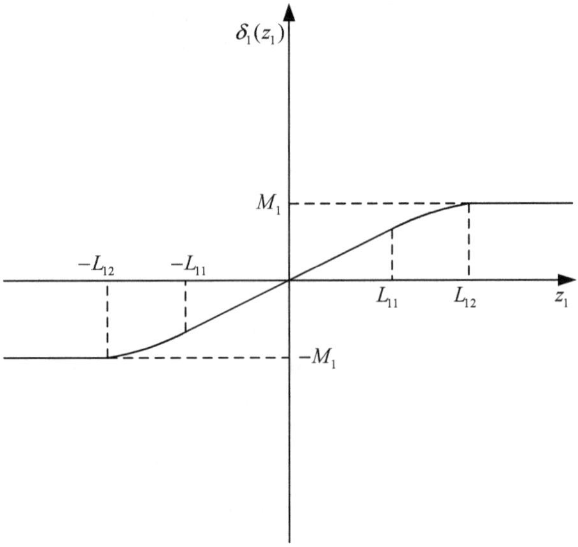

Fig. 5.8 $\delta_1(z_1)$: robust term of virtual control α_1

$$\alpha_1 = \alpha_{1a} + \alpha_{1s}, \tag{5.31}$$

$$\alpha_{1a} = \dot{y}_d, \ \alpha_{1s} = -\delta_1(z_1), \tag{5.32}$$

where $\delta_1(z_1)$ is designed to be a smooth and nondecreasing function, which has the following four properties:

- If $|z_1| \le L_{11}$, then $\delta_1(z_1) = k_{11}z_1$;
- $z_1\delta_1(z_1) > 0, \forall z_1 \neq 0$;
- $|\delta_1(z_1)| \le M_1, \forall z_1 \in \mathbb{R}$;
- $\left|\frac{\partial \delta_1}{\partial z_1}\right| \le k_{11}$, if $|z_1| \le L_{12}$, and $\left|\frac{\partial \delta_1}{\partial z_1}\right| = 0$, if $|z_1| \ge L_{12}$.

Specially, this function $\delta_1(z_1)$ can be drawn as in Fig. 5.8 and L_{11}, L_{12}, k_{11}, and M_1 are the positive design parameters to be chosen. Substituting (5.31)–(5.32) into (5.30) results in

$$\dot{z}_1 = z_2 - \delta_1(z_1), \tag{5.33}$$

which implies that $z_1 \to 0$ as $t \to \infty$, if $z_2 = 0$. Differentiating the error dynamics for $z_2 = x_2 - \alpha_1$ results in

$$m_s\dot{z}_2 = \psi(x, t) + \text{sat}(u) + F_l - m_s(\ddot{y}_d - \frac{\partial \delta_1}{\partial z_1}(z_2 - \delta_1)).$$

Finally, the control input u is designed as

$$u = u_a + u_s, \tag{5.34}$$
$$u_a = -\psi(x,t) + \hat{m}_s(\ddot{y}_d - \frac{\partial\delta_1}{\partial z_1}(z_2 - \delta_1)), \tag{5.35}$$
$$u_s = -\delta_2(z_2), \tag{5.36}$$

where $\hat{m}_s$ is the estimation of m_s, which is chosen as the projection type with the following form:

$$\dot{\hat{m}}_s = Proj_{m_s}(r_m^{-1}\tau), \tag{5.37}$$

$r_m > 0$ is a tunable gain and $\tau = (\ddot{y}_d - \frac{\partial\delta_1}{\partial z_1}(z_2 - \delta_1))z_2$. The standard projection mapping $Proj_{m_s}(r_m^{-1}\tau)$ is introduced as [4],

$$Proj_{m_s}(r_m^{-1}\tau) = \begin{cases} 0, \text{ if } \hat{m}_s = m_{s\max} \text{ and } r_m^{-1}\tau > 0, \\ 0, \text{ if } \hat{m}_s = m_{s\min} \text{ and } r_m^{-1}\tau < 0, \\ r_m^{-1}\tau, \text{ otherwise.} \end{cases}$$

$\delta_2(z_2)$ is shown in Fig. 5.9, and has the following properties:

- If $|z_2| \le L_{21}$, then $\delta_2(z_2) = k_{21}z_2$;
- If $L_{21} \le |z_2| \le L_{22}$, then $\delta_2(z_2) = k_{22}z_2$;
- $z_2\delta_2(z_2) > 0, \forall z_2 \ne 0$;
- $|\delta_2(z_2)| \le M_2, \forall z_2 \in \mathbb{R}$;
- $k_{21} \le k_{22}$ and $\left|\frac{\partial\delta_2}{\partial z_2}\right| = 0$, if $|z_2| \ge L_{22}$.

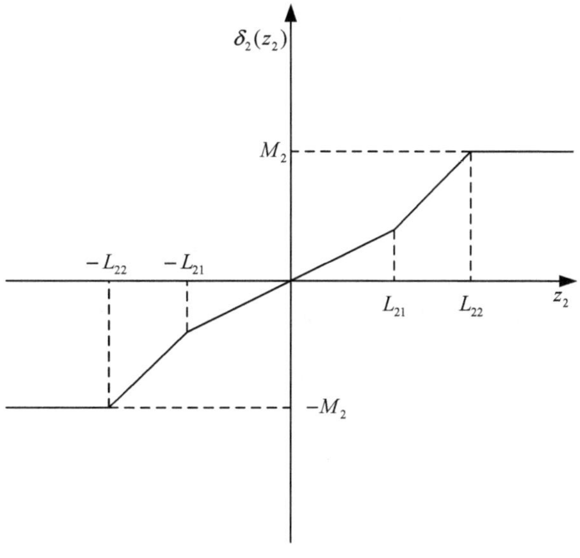

Fig. 5.9 $\delta_2(z_2)$: robust term of control input u

Performance Analysis of the Closed-loop Systems

Based on the above designing process, the following lemmas representing performance analysis of the closed-loop systems are given in this part.

Lemma 5.7 *If the control input u is designed as (5.34)–(5.36), then the set* $\Omega = \{z_1, z_2 : |z_1| \leq L_{11}, \ |z_2| \leq L_{22}\}$ *is a positive invariant set, as long as the following conditions are satisfied:*

1. $h + d < u_{\max}$,
2. $M_2 \geq 2u_{\max} - d$,
3. $|\ddot{y}_d| \leq \frac{-k_{11}(L_{22}+M_1)m_{s\max} - h - d + u_{\max} - \varepsilon}{m_{s\max}}$,
4. $k_{11}L_{11} > L_{22}$.

Proof Substituting the designed control input u into the error dynamics z_1 and z_2, the tracking error dynamics can be written as:

$$\begin{aligned}\dot{z}_1 &= z_2 - \delta_1(z_1),\\ m_s\dot{z}_2 &= \text{sat}\left(\hat{m}_s(\ddot{y}_d - \frac{\partial\delta_1}{\partial z_1}(z_2 - \delta_1)) - \psi(x,t) - \delta_2\right)\\ &\quad -m_s(\ddot{y}_d - \frac{\partial\delta_1}{\partial z_1}(z_2 - \delta_1)) + \psi(x,t) + F_l.\end{aligned}$$

In order to prove Ω is a positive invariant set, we need to prove that $z_1(t)$, $z_2(t)$ always stay in the set of Ω, as long as the initial values $z_1(0)$, $z_2(0)$ start inside Ω. Following [5–7], the proof can be divided into four cases:

Case 1: If z_2 hits the upper bound L_{22}, then according to Fig. 5.9, we have $\delta_2(z_2) = M_2$, which further implies that

$$\begin{aligned}u &= -\psi(x,t) + \hat{m}_s(\ddot{y}_d - \frac{\partial\delta_1}{\partial z_1}(z_2 - \delta_1)) - \delta_2(z_2)\\ &\leq h - M_2 + \hat{m}_s(|\ddot{y}_d| + \left|\frac{\partial\delta_1}{\partial z_1}\right|(|z_2| + |\delta_1|))\\ &\leq h - M_2 + \hat{m}_s(|\ddot{y}_d| + k_{11}(L_{22} + M_1)).\end{aligned}$$

Based on conditions 2 and 3 in Lemma 5.7, we have

$$\begin{aligned}u &\leq h - M_2 + u_{\max} - h - d - \varepsilon\\ &= u_{\max} - M_2 - d - \varepsilon\\ &\leq -u_{\max} - \varepsilon.\end{aligned}$$

Clearly, in this case, u is smaller than $-u_{\max}$, which means that the actuator output has been saturated, that is $\text{sat}(u) = -u_{\max}$. Therefore, the following equation is obtained:

$$\begin{aligned} m_s\dot{z}_2 &= -m_s(\ddot{y}_d - \frac{\partial \delta_1}{\partial z_1}(z_2 - \delta_1)) - u_{\max} + \psi(x,t) + F_l \\ &\leq -u_{\max} + h + d + m_x(|\ddot{y}_d| + k_{11}(L_{22} + M_1)) \\ &\leq -u_{\max} + h + d + u_{\max} - h - d - \varepsilon \\ &= -\varepsilon < 0. \end{aligned} \tag{5.38}$$

Inequality (5.38) tells us that when z_2 hits the upper bound L_{22}, z_2 will be decreased, and back to the set of Ω.

Case 2: Conversely, if z_2 hits the lower bound $-L_{22}$, then

$$\delta_2(z_2) = -M_2 \Longrightarrow u \geq u_{\max} + \varepsilon,$$

which implies $\text{sat}(u) = u_{\max}$. Then we have

$$\begin{aligned} m_s\dot{z}_2 &= u_{\max} - m_s(\ddot{y}_d - \frac{\partial \delta_1}{\partial z_1}(z_2 - \delta_1)) + \psi(x,t) + F_l \\ &\geq u_{\max} - h - d - (u_{\max} - h - d - \varepsilon) \\ &= \varepsilon > 0, \end{aligned} \tag{5.39}$$

which means that when z_2 hits the lower bound $-L_{22}$, z_2 will be increased, and back to the set of Ω as well.

Case 3: Similar to Case 1 and Case 2, if z_1 hits the upper bound L_{11}, we have

$$\dot{z}_1 = z_2 - \delta_1(z_1) \leq L_{22} - k_{11}L_{11}.$$

According to condition 4 in Lemma 1, the following inequality can be obtained

$$\dot{z}_1 \leq L_{22} - k_{11}L_{11} \leq 0,$$

which implies z_1 will be decreased back to the set of Ω.

Case 4: Finally, if z_1 hits the lower bound, that is $z_1 = L_{11}$, we can obtain

$$\dot{z}_1 \geq 0,$$

which implies that z_1 will be increased back to the set of Ω.

From cases 1–4, we know that once the error states $z_{1,}$ z_2 hit their bounds, they will be back into the the set of Ω, which means that the set

$$\Omega = \left\{z_{1,}\ z_2 : |z_1| \leq L_{11},\quad |z_2| \leq L_{22}\right\}$$

is a positive invariant set. The proof is finished.

Lemma 5.8 *If the control input u is designed as shown in (5.34)–(5.36), inside the invariant set Ω, the performance constraints (5.27) and (5.28) will never been violated, as long as the following conditions are satisfied,*

1. $x_{1\min} + L_{11} \leq y_d \leq x_{1\max} - L_{11}$,
2. $x_{2\min} + (L_{22} + k_{11}l_{11}) \leq \dot{y}_d \leq x_{2\max} - (L_{22} + k_{11}l_{11})$,
3. $x_{1\max} + \sqrt{q} \leq z_{\max}$, *with* $q = \max\{x_3^2(0) + x_4^2(0),\ \epsilon_2\}$,
4. $k_f(\sqrt{q} + d_{w1}) + b_f(\sqrt{q} + d_{w2}) \leq (m_{s\min} + m_u)g$,

where d_{w1} and d_{w2} are the upper bounds of the disturbances z_r and $\dot{z}_r$.

Proof Because $\Omega = \left\{z_{1,}\ z_2 : |z_1| \leq L_{11},\quad |z_2| \leq L_{22}\right\}$ is a positive invariant set, the following inequations is given:

$$-L_{11} \leq z_1 \leq L_{11},$$

which is equal to

$$y_d - L_{11} \leq x_1 \leq y_d + L_{11}.$$

According to condition 1 in Lemma 5.8, we have

$$x_{1\min} \leq x_1 \leq x_{1\max}.$$

Following a similar procedure, we have

$$x_{2\min} \leq x_2 \leq x_{2\max}.$$

The above design yields to a 2nd order error dynamic, while the original system is a 4th order system. So the zero dynamics consists of two states. We set $z_1 = 0$, which implies $\dot{z}_1 = z_2 = 0$. Hence, we obtain:

$$m_s\dot{z}_2 = \text{sat}(u) + \psi(x, t) + F_l - m_s\ddot{y}_d = 0, \tag{5.40}$$

and then

$$\begin{aligned}\text{sat}(u) &= -\psi(x, t) - F_l - m_s\ddot{y}_d \\ &\leq h + d + u_{\max} - h - d - k_{11}(L_{22} + M_1) - \varepsilon \\ &\leq u_{\max} - k_{11}(L_{22} + M_1) - \varepsilon. \end{aligned} \tag{5.41}$$

On the other side,

$$\begin{aligned}\text{sat}(u) &= -\psi(x,t) - F_l - m_s\ddot{y}_d \\ &\geq -(u_{\max} - h - d - k_{11}(L_{22} + M_1) - \varepsilon) - h - d \\ &\geq -u_{\max} + k_{11}(L_{22} + M_1) + \varepsilon.\end{aligned} \tag{5.42}$$

From Eqs. (5.41) and (5.42), we have

$$-u_{\max} < \text{sat}(u) < u_{\max},$$

which implies

$$\text{sat}(u) = -\psi(x,t) - F_l - m_s\ddot{y}_d. \tag{5.43}$$

If we use sat(u) in (5.43) to replace the one in $\dot{x}_4$, we obtain the following zero dynamics:

$$\dot{x} = Ax + \bar{w}, \tag{5.44}$$

where

$$x = \begin{bmatrix} x_3 \\ x_4 \end{bmatrix}, \ A = \begin{bmatrix} 0 & 1 \\ -\frac{k_t}{m_u} & -\frac{k_b}{m_u} \end{bmatrix},$$
$$\bar{w} = \begin{bmatrix} 0 \\ -m_s\ddot{y}_d - F_l + k_t z_r + k_b \dot{z}_r \end{bmatrix}.$$

Defining a positive function $V_0 = x^T x$, we have

$$\dot{V}_0 = x^T(A^T + A)x + 2x^T\bar{w}.$$

It is easy to verify that the matrix A has eigenvalues with negative real parts. Hence, we have $A^T + A < 0$. Noting that

$$2x^T\bar{w} \leq \frac{1}{\eta}x^T x + \eta\bar{w}^T\bar{w},$$

where η is a tuning positive value, and assuming $\bar{w}^T\bar{w}$ is bounded by ϵ_2, the following inequality is obtained.

$$\dot{V}_0 \leq [-\lambda_{\min}(A^T + A) + \frac{1}{\eta}]V_0 + \epsilon_2.$$

Based on properly chosen tuning value η, we can guarantee

$$-\lambda_{\min}(A^T + A) + \frac{1}{\eta} = -\epsilon_1,$$

where ϵ_1 is a positive value. Then,

$$\dot{V}_0 \leq -\epsilon_1 V + \epsilon_2. \tag{5.45}$$

Equation (5.45) shows that the Lyapunov function is bounded by

$$V_0(t) \leq V_0(0)e^{-\epsilon_1 t} + \epsilon_2(1 - e^{-\epsilon_1 t}) \leq \max\{V_0(0),\ \epsilon_2\} = q,$$

which tells us that $|x_k| \leq \sqrt{q}$, $k = 3, 4$.

From the above proof, we can see all the four states are constrained within their known bounds, and therefore, the performance constraints can be guaranteed by adjusting the bounds of the states as follows:

$$|x_1 - x_3| \leq x_{1\max} + \sqrt{q}, \tag{5.46}$$

if $x_{1\max} + \sqrt{q} \leq z_{\max}$, then the suspension space will be constrained within its range. Similarly, we have

$$|F_t + F_b| < k_f(\sqrt{q} + d_{w1}) + b_f(\sqrt{q} + d_{w2}). \tag{5.47}$$

If $k_f(\sqrt{q} + d_{w1}) + b_f(\sqrt{q} + d_{w2}) \leq (m_{s\min} + m_u)g$, then we can guarantee the constrained condition (5.27).

Lemma 5.9 *If the control input u is designed as shown in (5.34)–(5.36), inside the invariant set Ω, the steady-state output tracking error z_1 is bounded by*

$$|z_1(\infty)| \leq \frac{2u_{\max} - 2h - d - 2\varepsilon}{m_{s\min}k_{11}k_{22}},$$

and this steady-state output tracking error can be arbitrarily small, by adjusting feedback gains k_{11}, k_{22}.

Proof Define a positive function as

$$V_1 = \frac{m_s}{2}z_2^2. \tag{5.48}$$

If the actuator saturation doesn't occurred, that is to say $|u| \leq u_{\max}$, we have

$$
\begin{aligned}
\dot{V}_1 &= m_s z_2 \dot{z}_2 \\
&= z_2[-\psi(x,t) + \hat{m}_s(\ddot{y}_d - \frac{\partial \delta_1}{\partial z_1}(z_2 - \delta_1)) - \delta_2(z_2) \\
&\quad - m_s(\ddot{y}_d - \frac{\partial \delta_1}{\partial z_1}(z_2 - \delta_1))] + \psi(x,t) + F_l \\
&= z_2[\tilde{m}_s(\ddot{y}_d - \frac{\partial \delta_1}{\partial z_1}(z_2 - \delta_1)) + F_l - \delta_2(z_2)] \\
&\leq 2m_{s\max}(|\ddot{y}_d| + k_{11}(|z_1| + k_{11}L_{11}))] + |z_2|\,[d - k_{22}\,|z_2| \\
&\leq |z_2|\,[2u_{\max} - 2h - 2d - 2\varepsilon + d - k_{22}\,|z_2|] \\
&= -\frac{k_{22}}{2}[|z_2| - \frac{2u_{\max} - 2h - d - 2\varepsilon}{k_{22}}]^2 \\
&\quad + \frac{(2u_{\max} - 2h - d - 2\varepsilon)^2}{2k_{22}} - \frac{k_{22}}{2} z_2^2 \\
&\leq -\frac{k_{22}}{m_s} V_1 + \frac{(2u_{\max} - 2h - d - 2\varepsilon)^2}{2k_{22}},
\end{aligned}
$$

which implies that

$$
V_1(t) \leq \frac{(2u_{\max} - 2h - d - 2\varepsilon)^2}{2k_{22}}(1 - e^{-\frac{k_{22}}{m_s}t}) + V_1(0)e^{-\frac{k_{22}}{m_s}t},
$$

and then we have

$$
V_1(\infty) \leq \frac{(2u_{\max} - 2h - d - 2\varepsilon)^2}{2k_{22}},
$$

which means that

$$
z_2(\infty) \leq \frac{2u_{\max} - 2h - d - 2\varepsilon}{m_s k_{22}} \leq \frac{2u_{\max} - 2h - d - 2\varepsilon}{m_{s\min} k_{22}}.
$$

Because of $\dot{z}_1 = z_2 - k_{11} z_1$, we have

$$
z_2(\infty) - k_{11} z_1(\infty) = 0 \Longrightarrow z_1(\infty) \leq \frac{2u_{\max} - 2h - d - 2\varepsilon}{m_{s\min} k_{11} k_{22}}.
$$

If the ideal actuator output is smaller than the lower bound of saturation, that is $u < -u_{\max}$, we can obtain that

$$
\begin{aligned}
\dot{V}_1 &= m_s z_2 \dot{z}_2 \\
&= z_2[-u_{\max} + \psi(x,t) + F_l - m_s(\ddot{y}_d - \frac{\partial \delta_1}{\partial z_1}(z_2 - \delta_1))] \\
&\leq |z_2|\,[-u_{\max} + h + d + u_{\max} - h - d - \varepsilon] \\
&= -\,|z_2|\,\varepsilon \leq 0.
\end{aligned}
$$

Similarly, when $u > u_{\max}$, we have $\dot{V}_1 \le 0$. Finally, inside Ω, the steady-state output tracking error z_1 is bounded by

$$|z_1(\infty)| \le \frac{2u_{\max} - 2h - d - 2\varepsilon}{m_{s\,\min} k_{11} k_{22}},$$

and the proof is finished.

Lemma 5.10 *Suppose that u is given in (5.34)–(5.36). If after a finite time t_1, there exist parametric uncertainties only (i.e., $F_l = 0$, $\forall t \ge t_1$), then zero final output tracking error is also achieved, i.e, $z_1 \to 0$ as $t \to \infty$.*

Proof If the disturbances vanish after a finite time ($F_l = 0$), then define a positive function as

$$V_2 = \frac{m_s}{2} z_2^2 + \frac{1}{2} r_m^{-1} \tilde{m}_s^T \tilde{m}_s, \tag{5.49}$$

whose derivative is given as

$$\begin{aligned} \dot{V}_2 &= m_s z_2 \dot{z}_2 + r_m^{-1} \dot{\hat{m}}_s \tilde{m}_s \\ &= z_2[-m_s(\ddot{y}_d - \frac{\partial \delta_1}{\partial z_1}(z_2 - \delta_1)) + \text{sat}(u) + \psi(x,t)] \\ &\quad + r_m^{-1} \dot{\hat{m}}_s^T \tilde{m}_s. \end{aligned}$$

When actuator saturation doesn't occur, that is $|u| \le u_{\max}$, the following equation holds

$$\begin{aligned} \dot{V}_2 &= z_2[\tilde{m}_s(\ddot{y}_d - \frac{\partial \delta_1}{\partial z_1}(z_2 - \delta_1)) - \delta_2(z_2)] + r_m^{-1} \dot{\hat{m}}_s \tilde{m}_s \\ &= +\tilde{m}_s(r_m^{-1} \dot{\hat{m}}_s + (\ddot{y}_d - \frac{\partial \delta_1}{\partial z_1}(z_2 - \delta_1))z_2) - z_2 \delta_2(z_2). \end{aligned}$$

Noticing the property of the projection mapping

$$Proj_{m_s}(r_m^{-1}\tau) : \tilde{m}_s(r_m^{-1} Proj_{m_s}(r_m \tau) - \tau) \le 0, \forall \tau,$$

we have

$$\dot{V} \le -z_2 \delta_2(z_2) \le 0. \tag{5.50}$$

When actuator saturation happens, that is $u < -u_{\max}$ or $u > u_{\max}$, we have

$$\dot{V} \le -\varepsilon |z_2| \le 0. \tag{5.51}$$

Inequalities (5.50) and (5.51) imply that z_2 converges to zero, by using barbalat's Lemma, and thus we have $z_1 \to 0$, because of the stable transfer function

$$\frac{z_1(s)}{z_2(s)} = \frac{1}{s + k_{11}}.$$

The proof is finished.

Main Results

Theorem 5.11 *If u is designed as shown in (5.34)–(5.36), and the following conditions 1 – 8 hold, then the set*

$$\Omega = \left\{ z_1, z_2 : |z_1| \leq L_{11}, \;\; |z_2| \leq L_{22} \right\}$$

is a positive invariant set and inside Ω*, and the steady-state output tracking error* z_1 *is bounded by*

$$|z_1(\infty)| \leq \frac{2u_{\max} - 2h - d - 2\varepsilon}{m_{s\min} k_{11} k_{22}}.$$

At the same time, the performance constraints stated in (5.27)–(5.28) are never violated. Furthermore, if after a finite time t_1*, there exist parametric uncertainties only, then, zero final output tracking error is also achieved.*

1. $h + d < u_{\max}$,
2. $M_2 \geq 2u_{\max} - d$,
3. $|\ddot{y}_d| \leq \frac{-k_{11}(L_{22}+M_1)m_{s\max} - h - d + u_{\max} - \varepsilon}{m_{s\max}}$,
4. $k_{11} L_{11} > L_{22}$.
5. $x_{1\min} + L_{11} \leq y_d \leq x_{1\max} - L_{11}$,
6. $x_{2\min} + (L_{22} + k_{11} l_{11}) \leq \dot{y}_d \leq x_{2\max} - (L_{22} + k_{11} l_{11})$,
7. $x_{1\max} + \sqrt{q} \leq z_{\max}$,
8. $k_f(\sqrt{q} + d_{w1}) + b_f(\sqrt{q} + d_{w2}) \leq (m_{s\min} + m_u)g$.

Remark 5.12 As we state above, our main target is to stabilize the car body motion, and from Theorem 5.11, we can see that the vertical motion will be bounded within a infinitesimal range, as long as the initial errors $z_1(0)$, $z_2(0)$ are within the invariant set Ω. However, sometimes, the initial states are large, if absolute zero reference is assigned to the vertical motion of the vehicle body, the initial errors may be not in the set Ω. To handle this situation, a feasible approach is to choose the reference trajectory as

$$y_d(t) = \begin{cases} a_0 + a_1 t + a_2 t^2 + a_3 t^3 + a_4 t^4, & t < T_r \\ 0, & t \geq T_r \end{cases}, \tag{5.52}$$

where coefficient a_i, $i = 0, 1, 2, 3, 4$ are determined such that

$$y_d(0) = a_0 = x_1(0),$$
$$\dot{y}_d(0) = a_1 = x_2(0),$$
$$y_d(T_r) = \dot{y}_d(T_r) = \ddot{y}_d(T_r) = 0.$$

The above reference trajectory can guarantee that 1) $\dot{z}_1(0) = z_1(0) = 0$; 2) the vector function $y_d(t) \in C^3$. Furthermore, it is easy to see that $y_d(t) = 0$ and $\dot{y}_d(t) = 0$ can be reached in a pre-determined time T_r.

5.2.3 Comparative Experimental Results

In this section, an experimental plant is provided to illustrate the effectiveness of the proposed approach. The active suspension plant, as shown in Fig. 4.1, is a bench-scale model to emulate a quarter-car model, whose model parameters are listed in Table 5.3.

For this plant, all the initial state values are set as zeros, and we assume the initial mass $m_s = m_{s\min} = 2$ kg and the saturation limit $u_{\max} = 4$N. The controller parameters are given in Table 5.4.

The proposed controller is tested by the classic bump road input. Bump road inputs can be generally assumed as shocks. Shocks are discrete events of relatively short duration and high intensity, caused by, for example, a pronounced bump or pothole on an otherwise smooth road. The corresponding ground displacement is given by

Table 5.3 The model parameters of active suspension setup

Parameter	Value	Parameter	Value
$m_{s\min}$	2 kg	$m_{s\max}$	3 kg
k_{s1}	900 N/m	k_{n1}	10 N/m
b_e	8 Ns/m	b_c	7 Ns/m
k_f	2500 N/m	b_f	5 Ns/m

Table 5.4 The controller parameters of closed-loop system

Parameter	Value	Parameter	Value
r_m	100	k_{11}	100
k_{21}	150	k_{22}	100
L_{11}	0.01	L_{12}	0.02
L_{21}	0.01	L_{22}	0.02
M_1	1.6		

$$z_r = \begin{cases} \frac{h_0[1-\cos(8\pi t)]}{2}, & 1 \leq t \leq 1.25, \\ 0, & \text{otherwise}, \end{cases} \tag{5.53}$$

where h_0 is the height of the bump road input. It is assumed that the bump road input has the magnitude for $h_0 = 3$cm. Here, we assume the disturbance nonlinearity

$$F_l = \begin{cases} \sin(t), & 0 \leq t \leq 1, \\ 0, & \text{otherwise}. \end{cases}$$

For subsequent comparison, three cases are plotted:

S1: Passive suspension systems;
S2: Active suspension systems with standard ARC controller;
S3: Active suspension systems with the saturated ARC controller.

The responses of the passive suspension system, active suspension system with the standard adaptive robust controller and active suspension system with the proposed controller in (5.34)–(5.36), are compared in Fig. 5.10. In this figure, the plotted curves are the responses of cases S1–S3 which respect the cases of passive suspension system and active suspension system with the standard ARC controller and the proposed ARC controller. From the figure, we can see that the proposed controller can vanish the closed-loop system faster compared with the passive system, which clearly shows that an improved ride comfort has been achieved. Specially, there is nearly the same tracking performance between active suspension system with the proposed controller and active suspension system with the standard ARC controller, in spite of the presence of input saturation and performance constraits.

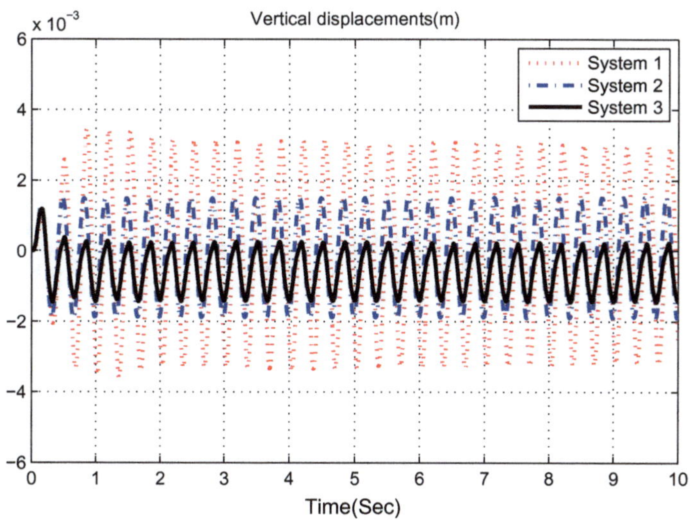

Fig. 5.10 Displacement responses of the vertical motion

Figure 5.11 are plotted to show the control input u.

In the active suspension control, the limitations of the suspension space should be taken into account, which means the suspension working space must be preserved. It can been observed from Fig. 5.12 that the controlled suspension spaces all fell into the acceptable ranges, whose maximums are $z_{\max} = 0.08$ m.

As stated in [8], to ensure a firm uninterrupted contact of wheels to road, the dynamic tire load should not exceed the static one. That is, the magnitude of $z_u - z_r$ should be as small as possible. The time responses of dynamic tire displacement $z_u - z_r$ for three cases are plotted in Fig. 5.13, which can be seen that the dynamic tire displacements are all within allowable ranges (the peak of the curves are 0.02m).

Besides, to investigate the effect of required settling time T_r on the system response, let the parameters of reference trajectory be $T_r = 1.0$s; 1.5s; 2.0s, respectively. Figures 5.14 and 5.15 show the time histories of vertical displacements and the corresponding control inputs for active suspensions with our proposed controller

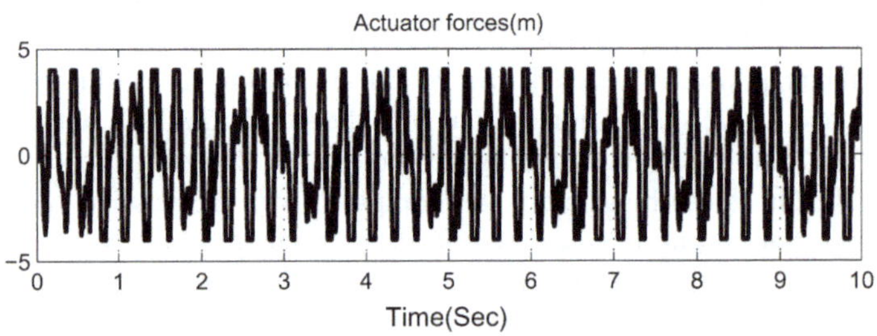

Fig. 5.11 Control inputs

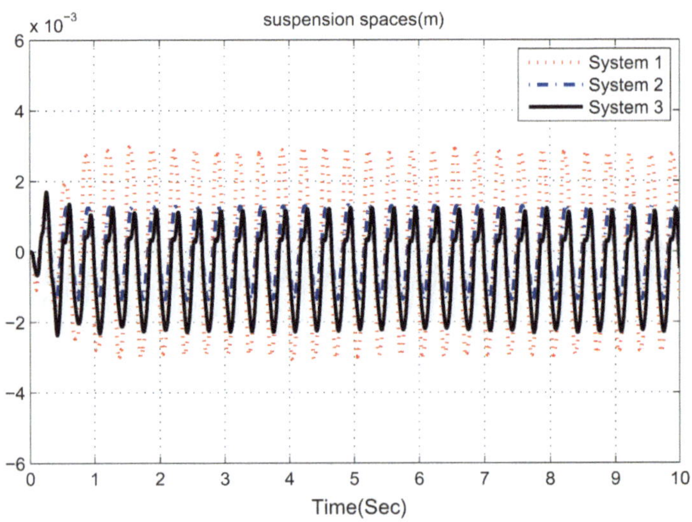

Fig. 5.12 Time responses of suspension spaces

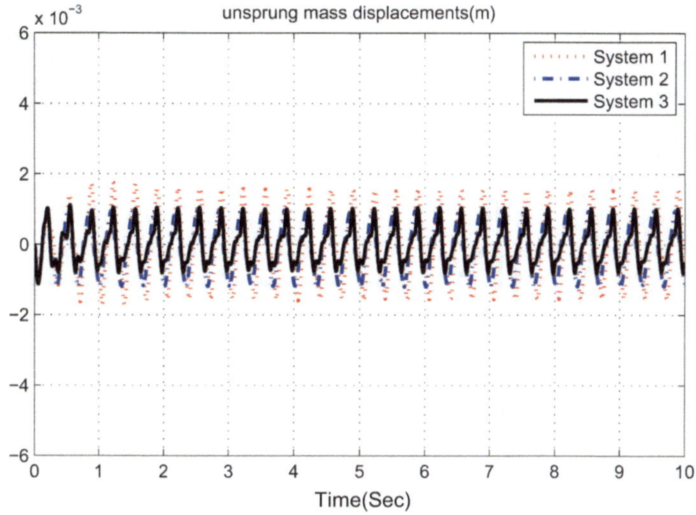

Fig. 5.13 Time responses of dynamic tire displacement $z_u - z_r$

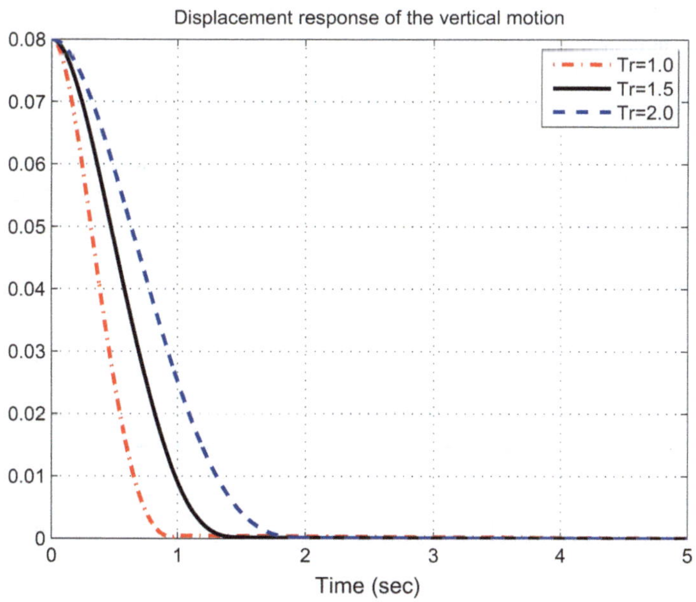

Fig. 5.14 Displacement responses of the vertical motion

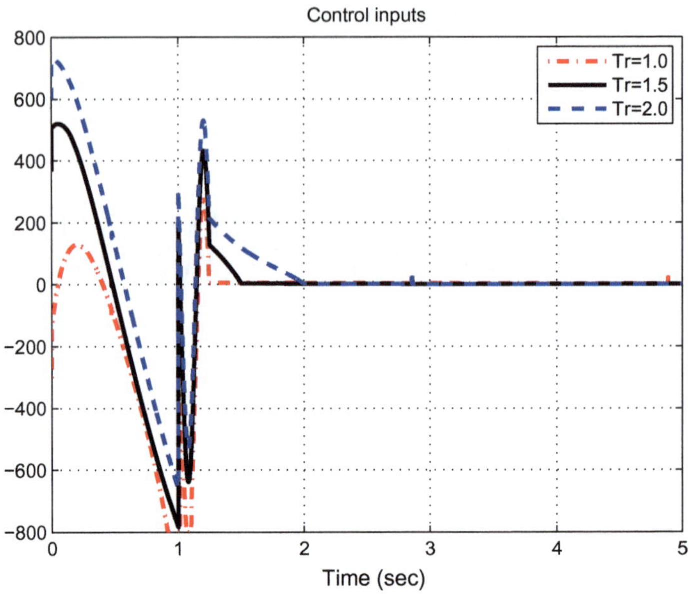

Fig. 5.15 Control inputs

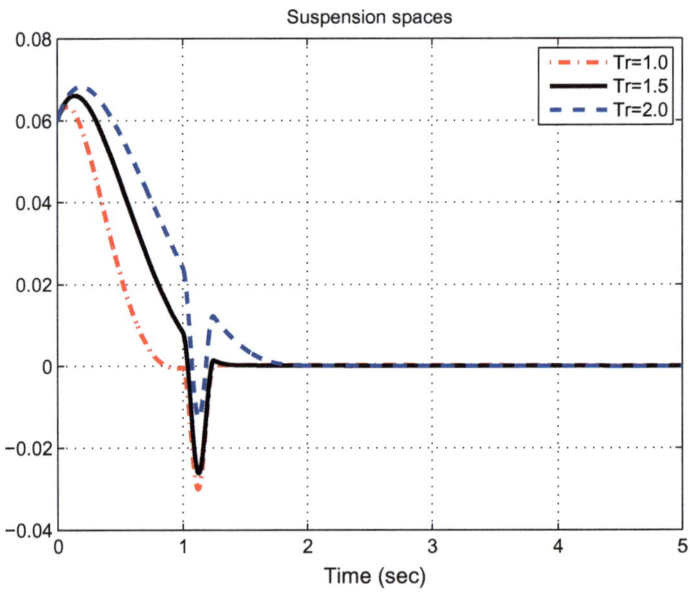

Fig. 5.16 Time responses of suspension spaces

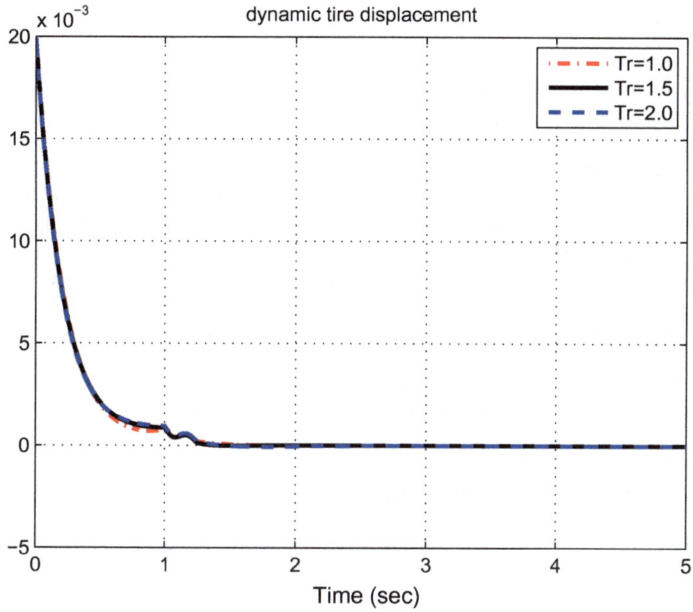

Fig. 5.17 Time responses of dynamic tire displacement $z_u - z_r$

in the case of the pre-determined settling time $T_r = 1.0$s; 1.5s; 2.0s, respectively, and Figs. 5.16 and 5.17 are the responses of suspension spaces and dynamic tire displacements. It can be seen that theoretically, we can settle T_r as arbitrarily value to make the vertical displacements converge to zero as we want. However, smaller settle time T_r results in larger control input force, which can be confirmed from Figs. 5.14 and 5.15.

5.2.4 Conclusion

In this section, an improved adaptive robust control strategy has been proposed for vehicle active suspension systems to stabilize the attitude of vehicle, where parameter uncertainties, external disturbances, actuator saturation and performance constraints are considered in a unified framework. Spring nonlinearity and the piece-wise linear behavior of the damper are taken into consideration to form the basis of accurate control. Furthermore, a reference trajectory is planned to keep the error initial values within the designed invariant set, and at the same time guarantee the vertical motion of car body stabilizing in pre-determined time. A nonlinear quarter-car model has been considered and the effectiveness of the proposed approach has been illustrated by a design example.

References

1. A. Teel, L. Zaccarian, J. Marcinkowski, An anti-windup strategy for active vibration isolation systems. Control Eng. Pract. **14**, 17–27 (2006)
2. N. Yagiz, Y. Hacioglu, Backstepping control of a vehicle with active suspensions. Control Eng. Pract. **16**, 1457–1467 (2008)
3. Y. Hong, B. Yao, A Globally stable high-performance adaptive robust control algorithm with Input saturation for percision motion control of linear motor drive systems. IEEE/ASME Trans. Mechtronics **12**(2), 198–207 (2007)
4. J. Yao, Z. Jiao, D. Ma, L. Yan, High-accuracy tracking control of hydraulic rotary actuators with modeling uncertainties. IEEE/ASME Trans. Mechatronics **19**(2), 633–641 (2014)
5. Y. Hong, B. Yao, A globally stable saturated desired compensation adaptive robust control for linear motor systems with comparative experiments. Automatica **43**(10), 1840–1848 (2007)
6. L. Lu, B. Yao, Globally stable fast tracking control of a chain of integrators with input saturation and disturbances: a holistic approach, in *Proceedings of the American Control Conference, San Francisco, CA, 2011*, pp. 4434–4439
7. L. Lu, B. Yao, W. Lin, A Two-loop Contour Tracking Control for Biaxial Servo Systems with Constraints and Uncertainties, in *Proceedings of the American Control Conference, Washington, DC, 2013*, pp. 6468–6473
8. W. Sun, H. Gao, O. Kaynak, Finite frequency H_∞ control for vehicle active suspension systems. IEEE Trans. Control Syst. Technol. **19**(2), 416–422 (2011)

Chapter 6
Active Suspension Control with the Unideal Actuators

The reliability of control strategies becomes more important in the case of active suspension control with the unideal actuators. In Sect. 6.1, the study of vehicle active suspension control with frequency band constraints and actuator input delay is carried out. Section 6.2 investigates the problem of robust sampled-data H_∞ control for active vehicle suspension systems. Finally, in Sect. 6.3, a fault tolerant control approach is proposed to deal with the problem of fault accommodation for unknown actuator failures of active suspension systems, where an adaptive robust controller is designed to adapt and compensate the parameter uncertainties, external disturbances and uncertain nonlinearities generated by the system itself and actuator failures.

6.1 Active Suspension Control with Frequency Band Constraints and Actuator Input Delay

This section investigates the problem of vehicle active suspension control with frequency band constraints and actuator input delay. Firstly, the mathematical model of suspension systems is established, and the problem of suspension control with finite frequency constraints is formulated to match the characteristics of the human body. Then, the finite frequency method is developed to deal with the problem of suspension control with actuator input delay, based on the generalized Kalman–Yakubovich–Popov (KYP) lemma. Compared with the traditional entire frequency approach for active suspension systems, the finite frequency approach proposed in this section achieves better disturbance attenuation performance for the chosen frequency range, and meantime the constraints required by real situation are guaranteed in the controller design. The effectiveness and merits of the proposed method are verified by a number of simulations with several types of road disturbances.

W. Sun et al., *Advanced Control for Vehicle Active Suspension Systems*, Studies in Systems, Decision and Control 204,
https://doi.org/10.1007/978-3-030-15785-2_6

6.1.1 Quarter-Car Model

By considering the vertical dynamics and taking into account the vehicle's symmetry, a suspension can be reduced to a quarter-car model is the first model in Sect. 1.2, and the state space expression is illustrated as (2.1) in Sect. 2. The performance requirements of the active suspensions are restated as follows.

Ride comfort

It is widely accepted that ride comfort is closely related to the body acceleration in frequency band 4–8 Hz. Consequently, it is important to keep the L_2 gain (from the disturbance inputs to car body acceleration) of closed-loop system as small as possible over the frequency band 4–8 Hz.

Road holding ability

Due to the disturbances caused by road bumpiness, a firm uninterrupted contact of wheels with road is important for vehicle handling and is essentially related to ride safety. Therefore, the dynamic tire load should be small, that is $k_t(z_u(t) - z_r(t)) < (m_s + m_u)g$.

Suspension deflection

To reduce the vertical acceleration of the car body, it is unavoidable to use more suspension travel, which increases the likelihood of a driver hitting the suspension travel limits when driving over a speed bump or into a pothole. Hence, the suspension deflection should travel within its allowable range, that is $|z_s(t) - z_u(t)| \leq z_{\max}$, where $z_{\max}$ is the maximum suspension deflection.

Actuator limitation

As the operator of a control system, the actuator plays an important role in engineering applications [1]. Here, another hard constraint imposed on active suspensions is from the limited power of the actuator, that is $|u(t)| \leq u_{\max}$.

In order to satisfy the above-mentioned performance requirements, the controlled outputs are defined by (2.7). Therefore, the state-space expression and deduction of the quarter-car model is derived as (2.8) and (2.9) in Chap. 2.

Time delays are widely encountered in the control loops because of the electrical and electromagnetic characteristics of the actuators. In this study, the results of finite frequency control are developed to address the active suspension systems with input time-delay.

We are interested in designing a state feedback controller

$$u(t) = Kx(t), \tag{6.1}$$

where K is the state feedback gain matrix to be designed. Therefore, the closed-loop system is given by:

$$
\begin{aligned}
\dot{x}(t) &= Ax(t) + BKx(t-d) + B_1 w(t), \\
z_1(t) &= C_1 x(t) + D_1 K x(t-d), \\
z_2(t) &= C_2 x(t).
\end{aligned} \tag{6.2}
$$

In this paper, our purpose is to design a state feedback gain matrix K such that

(1) the closed-loop system in (6.2) is asymptotically stable;

(2) the L_2 gain of closed-loop system should be smaller or less than a certain given value γ within the chosen frequency band, that is

$$
\|z_1(j\omega)\|_2 < \gamma \|w(j\omega)\|_2, \ \omega_1 < \omega < \omega_2; \tag{6.3}
$$

(3) the following constraints are guaranteed with the disturbance energy under the bound $w_{\max}$.

$$
\begin{aligned}
|\{z_2(t)\}_i| &\le 1, \ i = 1, 2, \\
|u(t)| &\le u_{\max}.
\end{aligned} \tag{6.4}
$$

6.1.2 Finite Frequency Controller Design

In this section, the finite frequency controller will be designed to address the active suspension systems with actuator input delay, and the following theorem gives the conclusion of controller design:

Theorem 6.1 *Give positive scalars γ, α, β_1, β_2, ρ and let a state feedback controller in the form of (6.1) be given. The closed-loop system in (6.2) is asymptotically stable, and satisfies $\|z_1(j\omega)\|_2 < \gamma \|w(j\omega)\|_2$, for $\omega \in [\omega_1, \omega_2]$, while respecting the constraints in (6.4) with the disturbance energy under the bound $w_{\max} = (\rho - V'(0))/\eta$, if there exist symmetric matrices P, $S_1 > 0$, $S_2 > 0$, $R_1 > 0$, $R_2 > 0$, $P_1 > 0$, $P_2 > 0$, $Q > 0$ and general matrices K and Y satisfying*

$$
\Pi_1 + [Y_1 U_1]_s < 0, \tag{6.5}
$$

$$
\Theta + M + [Y_2 U_2]_s < 0, \tag{6.6}
$$

$$
\begin{bmatrix} -I & \sqrt{\rho}\{C_2\}_i \\ * & -P_2 \end{bmatrix} < 0, \tag{6.7}
$$

$$
\begin{bmatrix} -I & \sqrt{\rho}K \\ * & -u_{\max}^2 P_2 \end{bmatrix} < 0, \tag{6.8}
$$

where

$$\begin{aligned}
\Pi_1 &= \begin{bmatrix} d^2 S_1 & P_1 & 0 \\ * & R_1 - S_1 & S_1 \\ * & * & -R_1 - S_1 \end{bmatrix}, \\
\Pi_2 &= \begin{bmatrix} d^2 S_2 & P_2 & 0 \\ * & R_2 - S_2 & S_2 \\ * & * & -R_2 - S_2 \end{bmatrix}, \\
Y_1 &= \begin{bmatrix} Y^T \ \alpha Y^T \ 0 \end{bmatrix}^T, \\
Y_2 &= \begin{bmatrix} \beta_1 Y^T \ \beta_2 Y^T \ 0 \ 0 \end{bmatrix}^T, \\
U_1 &= \begin{bmatrix} -I \ A \ BK \end{bmatrix}, \\
U_2 &= \begin{bmatrix} -I \ A \ BK \ B_1 \end{bmatrix}, \\
L &= \begin{bmatrix} 0 \ C_1 \ D_1 K \end{bmatrix}, \\
\Theta &= \begin{bmatrix} \Pi_2 + L^T L & 0 \\ * & -\gamma^2 I \end{bmatrix}, \\
F &= \begin{bmatrix} I \ 0 \ 0 \ 0 \\ 0 \ I \ 0 \ 0 \end{bmatrix}, \ \Phi = \begin{bmatrix} 0 & 1 \\ 1 & 0 \end{bmatrix}, \\
M &= F^*(\Phi \otimes P + \Psi \otimes Q)F, \\
\Psi &= \begin{bmatrix} -1 & j\omega_c \\ -j\omega_c & -\omega_1\omega_2 \end{bmatrix}.
\end{aligned}$$

Proof Firstly, the asymptotic stability of (6.2) with $w(t) = 0$ is shown, that is

$$\dot{x}(t) = Ax(t) + BKx(t - d). \tag{6.9}$$

Consider a Lyapunov functional candidate as

$$V(t) \triangleq V_1(t) + V_2(t) + V_3(t), \tag{6.10}$$

$$V_1(t) \triangleq x^T(t)P_1x(t), \tag{6.11}$$

$$V_2(t) \triangleq \int_{t-d}^{t} x^T(s)R_1x(s)ds, \tag{6.12}$$

$$V_3(t) \triangleq d\int_{-d}^{0}\int_{t+\beta}^{t} \dot{x}^T(\alpha)S_1\dot{x}(\alpha)d\alpha d\beta, \tag{6.13}$$

where $P_1 > 0,\ R_1 > 0$ and $S_1 > 0$ are matrices to be determined.

The derivatives of $V_1(t),\ V_2(t)$ and $V_3(t)$ satisfy

$$\begin{aligned}
\dot{V}_1(t) &= \dot{x}^T(t)P_1x(t) + x^T(t)P_1\dot{x}(t), \\
\dot{V}_2(t) &= x^T(t)R_1x(t) - x^T(t-d)R_1x(t-d), \\
\dot{V}_3(t) &= d^2\dot{x}^T(t)S_1\dot{x}(t) - d\int_{t-d}^{t} \dot{x}^T(\beta)S_1\dot{x}(\beta)d\beta.
\end{aligned}$$

Lemma 6.2 (Jensen inequality) *For any positive symmetric constant matrix* $M \in R^{n\times n}$, *scalar* r *satisfying* $r > 0$, *and a vector function* $\omega : [0, r] \longrightarrow R_n$ *such that the integrations concerned are well defined, then*

$$r\int_0^r \omega^T(s)M\omega(s)d\beta \geq \left(\int_0^r \omega(s)ds\right)^T M\int_0^r \omega(s)ds.$$

By using Jensen inequality in Lemma 6.2, we have

$$\begin{aligned}&-d\int_{t-d}^{t} \dot{x}^T(\beta)S_1\dot{x}(\beta)d\beta\\ &\leq -\left[x(t)-x(t-d)\right]^T S_1\left[x(t)-x(t-d)\right].\end{aligned}$$

Then, we have $\dot{V}(t) \leq \zeta^T(t)\Pi_1\zeta(t)$, where $\zeta(t) = \left[\dot{x}^T(t)\ x^T(t)\ x^T(t-d)\right]^T$. On the other hand, from Lemma 6.3,

Lemma 6.3 (Finsler's Lemma) *Let* $x \in R_n$, $P \in S_n$, *and* $H \in R^{m\times n}$ *such that* $\text{rank}(H) = r < n$. *The following statements are equivalent:*

$$\begin{aligned}&x^T P x < 0, \forall Hx = 0, x \neq 0\\ &\Leftrightarrow \exists X \in R^{n\times m} : P + XH + H^T X^T < 0.\end{aligned}$$

inequality (6.5) is equivalent to

$$\delta^T(t)\Pi_1\delta(t) < 0,\ \ \forall U_1\delta(t) = 0,$$

which can guarantee $\dot{V}(t) < 0$ from the fact that $U_1\zeta(t) = 0$, that means the closed-loop system in (6.2) is asymptotically stable.

Next, we shall establish the L_2 gain performance of closed-loop system in (6.3). Choose a Lyapunov functional as

$$V'(t) \triangleq V_1'(t) + V_2'(t) + V_3'(t), \tag{6.14}$$

$$V_1'(t) \triangleq x^T(t)P_2x(t), \tag{6.15}$$

$$V_2'(t) \triangleq \int_{t-d}^{t} x^T(s)R_2x(s)ds, \tag{6.16}$$

$$V_3'(t) \triangleq d\int_{-d}^{0}\int_{t+\beta}^{t} \dot{x}^T(\alpha)S_2\dot{x}(\alpha)d\alpha d\beta, \tag{6.17}$$

where $P_2 > 0,\ R_2 > 0$ and $S_2 > 0$ are matrices to be determined. Then, we can obtain

$$\dot{V}'(t) \leq \zeta^T(t)\Pi_2\zeta(t). \tag{6.18}$$

Define

$$J \triangleq \|z_1\|_2^2 - \gamma^2 \|w\|_2^2 . \tag{6.19}$$

Under zero initial conditions, we can obtain

$$\begin{aligned} J &\le \|z_1\|_2^2 - \gamma^2 \|w\|_2^2 + V'(\infty) - V'(0) \\ &= \int_0^\infty (z_1^T z_1 - \gamma^2 w^T w + \dot{V}'(t))dt \qquad (6.20) \\ &\le \int_0^\infty \xi^T(t)\Theta\xi(t)dt, \qquad (6.21) \end{aligned}$$

where $\xi(t) \triangleq \left[\zeta^T(t)\ w^T(t)\right]^T$. Define

$$\bar{J} = \int_0^\infty \xi^T(t)\Theta\xi(t)dt. \tag{6.22}$$

Noting that Θ is a real symmetric matrix, we can split Θ as $\Theta = (\Theta^{\frac{1}{2}})^*\Theta^{\frac{1}{2}}$, and we can get

$$\bar{J} = \int_0^\infty \phi^*(t)\phi(t)dt, \text{ with } \phi(t) = \Theta^{\frac{1}{2}}\xi(t). \tag{6.23}$$

After Fourier transform to $\phi(t)$, we can obtain the spectrum of $\phi(t)$, which is denoted as $\phi_s(\omega)$. By using Parseval equality, we have

$$\begin{aligned} \bar{J} &= \int_0^\infty \phi^*(t)\phi(t)dt = \frac{1}{2\pi}\int_{-\infty}^\infty \phi_s^*(\omega)\phi_s(\omega)d\omega \\ &= \frac{1}{2\pi}\int_{-\infty}^\infty \xi_s^*(\omega)\Theta\xi_s(\omega)d\omega. \qquad (6.24) \end{aligned}$$

On the other hand, Lemma 6.3 tells us that inequality (6.6) is equivalent to

$$\begin{aligned} \xi_s^*(\omega)(\Theta + M)\xi_s(\omega) &< 0, \qquad (6.25) \\ U_2\xi_s(\omega) &= 0, \end{aligned}$$

where inequality (6.25) can guarantee

$$\xi_s^*(\omega)\Theta\xi_s(\omega) < 0, \text{ with } \xi_s^*(\omega)M\xi_s(\omega) \ge 0, \tag{6.26}$$

by using S-procedure.

Define

$$\begin{aligned} \mathbf{M} &= \left\{F^*(\Phi \otimes P + \Psi \otimes Q)F;\ P,\ Q \in \mathbf{H}_2\right\}, \\ \mathbf{E} &= \{s = \mathbb{C} : v(s,\Phi) = 0, v(s,\Psi) \ge 0\}. \end{aligned}$$

From [2], we know the following two sets are equivalent:

$$\mathbf{W}_{(1)} = \left\{\epsilon \in \mathbb{C} : \epsilon \neq 0,\ \epsilon^* M \epsilon \geq 0,\ \exists M \in \mathbf{M}\right\},$$
$$\mathbf{W}_{(2)} = \{\epsilon \in \mathbb{C} : \epsilon \neq 0,\ T_s F \epsilon = 0,\ s \in \mathbf{E}\},$$

and $\mathbf{W}_{(2)}$ describes our chosen frequency band $\omega \in \left[\omega_1, \omega_2\right]$, with

$$\Phi = \begin{bmatrix} 0 & 1 \\ 1 & 0 \end{bmatrix},\ \Psi = \begin{bmatrix} -1 & j\omega_c \\ -j\omega_c & -\omega_1\omega_2 \end{bmatrix}.$$

Therefore, we can see inequality (6.26) guarantee $J < 0$, which implies $\|z_1(j\omega)\|_2 < \gamma \|w(j\omega)\|_2$ with $\omega \in \left[\omega_1, \omega_2\right]$. Inequality (6.3) is guaranteed.

From the above proof, we can see inequality (6.26) can guarantee $\bar{J} < 0$, which implies

$$z_1^T z_1 - \gamma^2 w^T w + \dot{V}_1'(t) < 0, \tag{6.27}$$

and then inequality (6.27) guarantees

$$\dot{V}_1'(t) < \gamma^2 w^T w. \tag{6.28}$$

Integrating both sides of inequality (6.28) from 0 to t results in

$$V'(t) - V'(0) < \gamma^2 \int_0^t w^T(t) w(t) dt \leq \gamma^2 w_{\max},$$

where $w_{\max} = \|w\|_2^2$. Note that $V_2'(t) > 0$, $V_3'(t) > 0$, which shows that

$$x^T(t) P_2 x(t) < V'(0) + \gamma^2 w_{\max} = \rho. \tag{6.29}$$

Consider

$$\max_{t\geq 0} |\{z_2(t)\}_i|^2 = \max_{t\geq 0}(x^T(t)\{C_2\}_i^T \{C_2\}_i x(t)),$$
$$\max_{t\geq 0} |u(t)|^2 = \max_{t\geq 0}(x^T(t) K^T K x(t)).$$

From inequality (6.29), it is true that

$$\max_{t\geq 0} |\{z_2(t)\}_i|^2 < \rho \cdot \lambda_{\max}(P_2^{-\frac{1}{2}} \{C_2\}_i^T \{C_2\}_i P_2^{-\frac{1}{2}}),$$
$$\max_{t\geq 0} |u(t)|^2 < \rho \cdot \lambda_{\max}(P_2^{-\frac{1}{2}} K^T K P_2^{-\frac{1}{2}}),$$

where $\lambda_{\max}(\cdot)$ represents the maximum eigenvalue. Then, the constraints in (6.4) hold if

$$\rho P_2^{-\frac{1}{2}} \{C_2\}_i^T \{C_2\}_i P_2^{-\frac{1}{2}} < I, \; i = 1, 2,$$
$$\rho P_2^{-\frac{1}{2}} K^T K P_2^{-\frac{1}{2}} < u_{\max}^2 I, \tag{6.30}$$

which, by Schur complement, are equivalent to (6.8). The proof is completed. □

Define $\hat{J}_1 = \text{diag}\{Y^{-1}, \; Y^{-1}, \; Y^{-1}\}$, $\hat{J}_2 = \text{diag}\{Y^{-1}, \; Y^{-1}, \; Y^{-1}, \; I, \; I\}$, $\hat{J}_3 = \text{diag}\{I, Y^{-1}\}$, $\hat{J}_4 = \text{diag}\{I, Y^{-1}\}$. Then, we perform a congruence transformation to (6.5)–(6.8), respectively, by the full rank matrix $\hat{J}_1$, $\hat{J}_2$, $\hat{J}_3$ and $\hat{J}_4$ on the left, and $\hat{J}_1^T$, $\hat{J}_2^T$, $\hat{J}_3^T$ and $\hat{J}_4^T$ on the right. Defining

$$\begin{bmatrix} d^2\bar{S}_1 - [\bar{Y}]_s & \bar{P}_1 + A\bar{Y}^T - \alpha\bar{Y} & B\bar{K} \\ * & \bar{R}_1 - \bar{S}_1 + [\alpha A\bar{Y}^T]_s & \bar{S}_1 + \alpha B\bar{K} \\ * & * & -\bar{R}_1 - \bar{S}_1 \end{bmatrix} < 0, \tag{6.31}$$

$$\begin{bmatrix} \begin{pmatrix} d^2\bar{S}_2 - \bar{Q} \\ -\beta_1 [\bar{Y}]_s \end{pmatrix} & \begin{pmatrix} \bar{P}_2 + \bar{P} + j\omega_c\bar{Q} \\ +\beta_1 A\bar{Y}^T - \beta_2\bar{Y} \end{pmatrix} & \beta_1 B\bar{K} & \beta_1 B_1 & 0 \\ * & \begin{pmatrix} \bar{R}_2 - \bar{S}_2 - \omega_1\omega_2\bar{Q} \\ +\beta_2 [A\bar{Y}^T]_s \end{pmatrix} & \beta_2 B\bar{K} + \bar{S}_2 & \beta_2 B_1 & \bar{Y}C^T \\ * & * & -\bar{R}_2 - \bar{S}_2 & 0 & \bar{K}^T D_1^T \\ * & * & * & -\gamma^2 I & 0 \\ * & * & * & * & -I \end{bmatrix} < 0, \tag{6.32}$$

$$\bar{S}_1 = Y^{-1}S_1Y^{-T}, \bar{P}_1 = Y^{-1}P_1Y^{-T}, \bar{Y} = Y^{-1},$$
$$\bar{R}_1 = Y^{-1}R_1Y^{-T}, \bar{R}_2 = Y^{-1}R_1Y^{-T}, \; \bar{K} = KY^{-T},$$
$$\bar{S}_2 = Y^{-1}S_1Y^{-T}, \; \bar{P}_2 = Y^{-1}P_1Y^{-T},$$

we have the following solvable theorem.

Theorem 6.4 *Give positive scalars $\gamma, \alpha, \beta_1, \beta_2, \rho$ and let a state feedback controller in the form of (6.1) be given. The closed-loop system in (6.2) is asymptotically stable, and satisfies $\|z_1(j\omega)\|_2 < \gamma \|w(j\omega)\|_2$, for $\omega \in [\omega_1, \omega_2]$, while respecting the constraints in (6.4) with the disturbance energy under the bound $w_{\max} = (\rho - V'(0))/\eta$, if there exist symmetric matrices $\bar{P}$, $\bar{S}_1 > 0$, $\bar{S}_2 > 0$, $\bar{P}_1 > 0$, $\bar{P}_2 > 0$, $\bar{R}_1 > 0$, $\bar{R}_2 > 0$, $\bar{Q} > 0$ and general matrices $\bar{Y}$, $\bar{K}$, satisfying inequalities (6.31)–(6.34).*

$$\begin{bmatrix} -I & \sqrt{\rho}\{C_2\}_i \bar{Y}^T \\ * & -\bar{P}_2 \end{bmatrix} < 0, \tag{6.33}$$

$$\begin{bmatrix} -I & \sqrt{\rho}\bar{K} \\ * & -u_{\max}^2 \bar{P}_2 \end{bmatrix} < 0. \tag{6.34}$$

Moreover, the control gain K is given by $K = \bar{K}\bar{Y}^{-T}$.

6.1.3 Simulation Verification

In this section, we will apply the above approach to design a state feedback controller based on the quarter-car model described in Sect. 2.1.2. The quarter-car model parameters are shown as: $m_s = 320\,\text{kg}$, $m_u = 40\,\text{kg}$, $k_s = 18\,\text{kN/m}$, $k_t = 200\,\text{kN/m}$, $c_s = 1\,\text{kNs/m}$, $c_t = 10\,\text{Ns/m}$.

By solving the matrix inequalities (6.31)–(6.34) with $\omega_1 = 4\,\text{Hz}$, $\omega_2 = 8\,\text{Hz}$, $\rho = 0.01$, $z_{\max} = 100\,\text{mm}$ and choosing $d = 5\,\text{ms}$, we can obtain $\gamma_{\min} = 8.4059$, and

$$K = 10^4 \times \left[\, 1.6985\ 0.5127\ 0.0180\ -0.0654 \,\right].$$

Then, we will solve the entire frequency controller, according to [3]. After solving the matrix inequalities in [3] with $d = 5$ ms, we can calculate $\gamma_{\min} = 16.1799$, and

$$K = 10^4 \times \left[\, 1.7799\ 0.2873\ 0.0485\ -0.0308 \,\right].$$

After obtaining the finite frequency controller and the entire frequency controller, we will compare the two controllers to illustrate the performance of the closed-loop suspension system with actuator time delay in finite frequency domain. In Fig. 6.1, the solid and dotted lines are the responses of the closed-loop system with the finite frequency controller and the entire frequency controller, respectively, and the dashed line is the response of the passive system. From the figure, we can see that the finite frequency controller yields the least MSV (maximum singular values) over the frequency range 4–8Hz, for the active suspension systems with input delay ($d = 5$ ms), which clearly shows that an improved ride comfort has been achieved.

In order to evaluate the suspension characteristics with respect to the performance requirements, we give the disturbance signal in (6.35) to clarify the effectiveness of our finite frequency controller.

$$w(t) = \begin{cases} A\sin(2\pi f t), & \text{if } 0 \le t \le T, \\ 0, & \text{if } t > T. \end{cases} \tag{6.35}$$

Assume $A = 0.5\,\text{m}$, $f = 5\,\text{Hz}$ and $T = 1/f = 0.2\,\text{s}$, and the time-domain response of body vertical acceleration for the active suspension system is shown in Fig. 6.2, where the black solid line and the blue dashed line are the responses of body vertical acceleration with the finite frequency controller and the entire frequency controller, respectively. We can clearly see that the value of the body acceleration with the finite frequency controller is less than that with the entire frequency controller. In addition, Fig. 6.3 shows that the ratio $x_1(t)/z_{\max}$ and the relative dynamic tire load $k_t x_2(t)/(m_s + m_u)g$ are below 1, which means the time-domain constraints are guaranteed by the designed controller. Also, the force of actuator is below the maximum 2500 N.

When the actuator time delay is increased to 20 ms, the maximum singular values of the passive suspension, the active suspension with entire frequency controller and

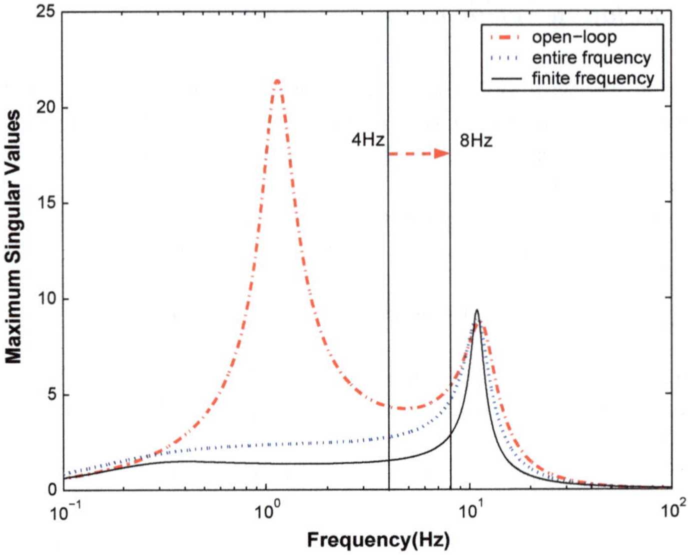

Fig. 6.1 The curves of maximum singular values ($d = 5$ ms)

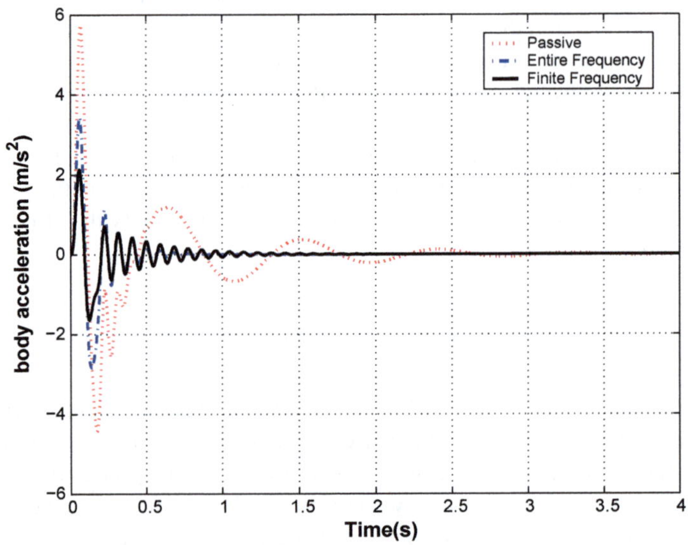

Fig. 6.2 The time-domain response of body acceleration ($d = 5$ ms)

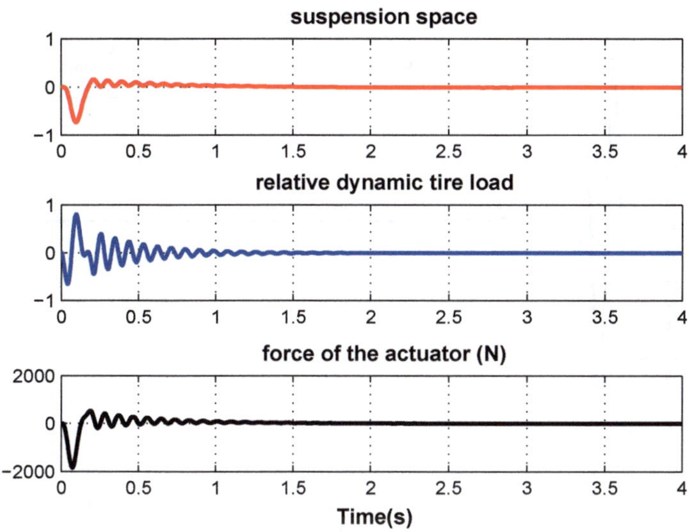

Fig. 6.3 The constraints of suspension system ($d = 5$ ms)

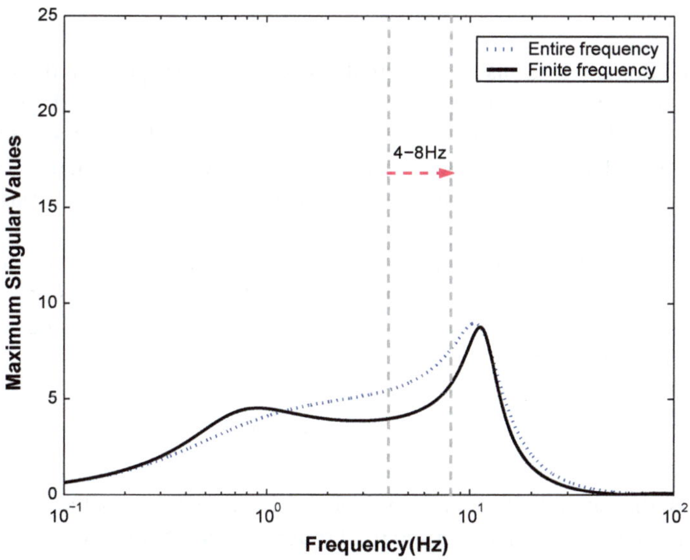

Fig. 6.4 The curves of maximum singular values with ($d = 20$ ms)

the active suspension with finite frequency controller are compared in Fig. 6.4, which shows the same conclusion with the case $d = 5$ ms. However, as the actuator delay increases, disturbance attenuation becomes weaker than the case of $d = 5$ ms, which also implies the impact of actuator time delay on the suspension system.

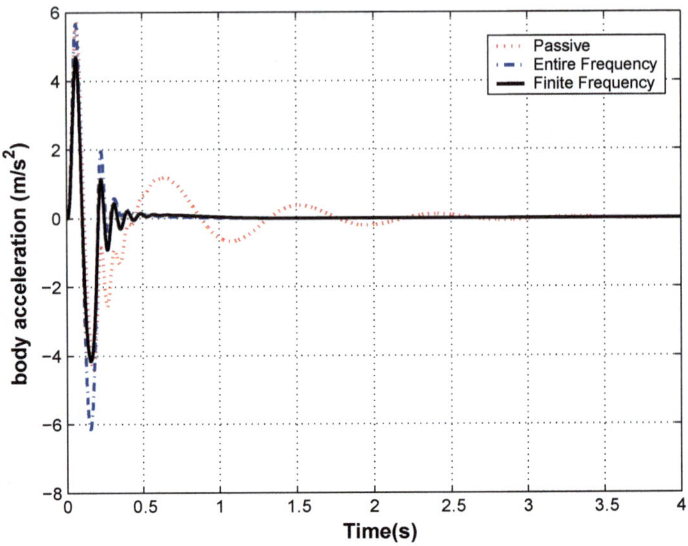

Fig. 6.5 The time-domain response of body acceleration ($d = 20$ ms)

The time-domain responses of the body acceleration, suspension deflection, relative tire dynamic load, and active force are plotted in Figs. 6.5 and 6.6. It can be seen from these figures that the responses of the body acceleration, the suspension deflection, the relative tire dynamic load, and the active force of active suspension are all similar to those shown in Figs. 6.2 and 6.3 in spite of the increase of time delay.

Hereafter, another disturbance signal is used to verify the effectiveness of the designed controller, that is the random vibration. Random vibrations are consistent and typically specified as random process with a given ground displacement power spectral density (PSD) of

$$G_q(n) = G_q(n_0)(\frac{n}{n_0})^{-W}, \tag{6.36}$$

where n is the spatial frequency and n_0 is the reference spatial frequency of $n_0 = 0.1(1/\text{m})$; $G_q(n_0)$ stands for the road roughness coefficient; $W = 2$ is the road roughness constant. PSD ground velocity is given by

$$G_{\dot{q}}(f) = (2\pi f)^2 G_q(f) = 4\pi G_q(n_0) n_0^2 V, \tag{6.37}$$

which is only related with the vehicle forward velocity, and the process of the mathematical derivation is stated in Sect. 1.3. Select the road roughness as $G_q(n_0) = 256 \times 10^{-6}\,\text{m}^3$, which is corresponded to D Grade (Poor) according to ISO2361 standards, to generate the random road profile. Set the vehicle forward velocity as

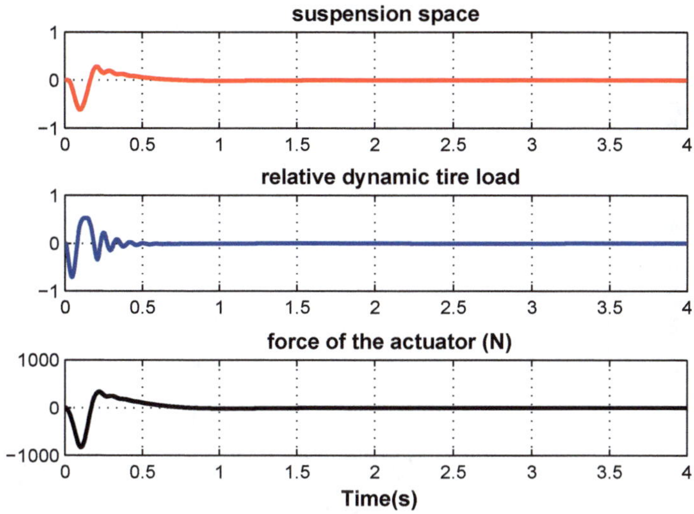

Fig. 6.6 The constraints of suspension system ($d = 20$ ms)

V = 45 km/h, and as expected, it is observed from Fig. 6.7 that the closed-loop system with finite frequency controller realizes a better ride comfort, compared with system with entire frequency controller for the frequency range 4–8 Hz (since the closed-loop system with finite frequency controller has lower PSD body acceleration than that with entire frequency controller, and smaller PSD body acceleration value results in better ride comfort), where PSD body acceleration can be calculated by

$$G_{z_1}(f) = |G(j\omega)|^2 G_{\dot{q}}(f). \tag{6.38}$$

To check more random road profiles, we select the road roughness as $G_q(n_0) = 16 \times 10^{-6}\,\mathrm{m}^3$ (B Grade, Good), $G_q(n_0) = \times 10^{-6}\,\mathrm{m}^3$ (C Grade, Average), and $G_q(n_0) = 1024 \times 10^{-6}\,\mathrm{m}^3$ (E Grade, Very Poor), respectively. From Fig. 6.8, it can be observed that the closed-loop system with finite frequency controller realizes a better ride comfort than that with the traditional method in spite of the different road roughness. When the actuator time delay is not 5 ms, but the other value, the control results are with the similar situation, and here we passes the repetition over.

6.1.4 Conclusion

In this section, the finite frequency method has been developed to deal with the active suspension systems with actuator input delay, and the state feedback controller for active suspension systems with frequency band constraints has been designed to improve ride comfort. The key idea of designing the proposed controllers is to use

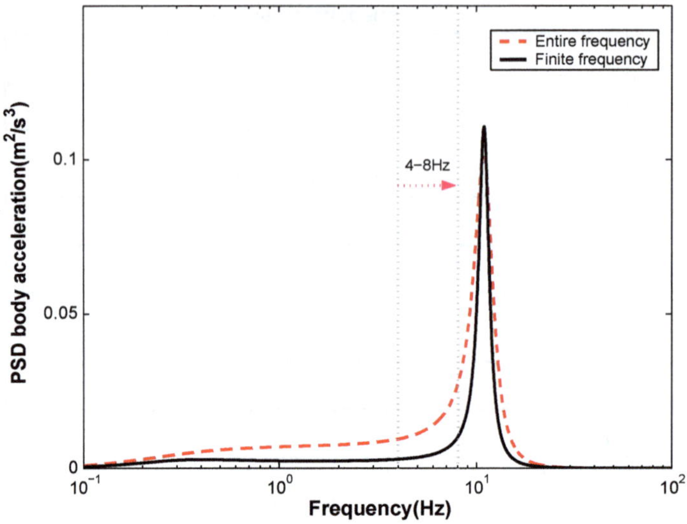

Fig. 6.7 The power spectral density of body acceleration ($d = 5$ ms)

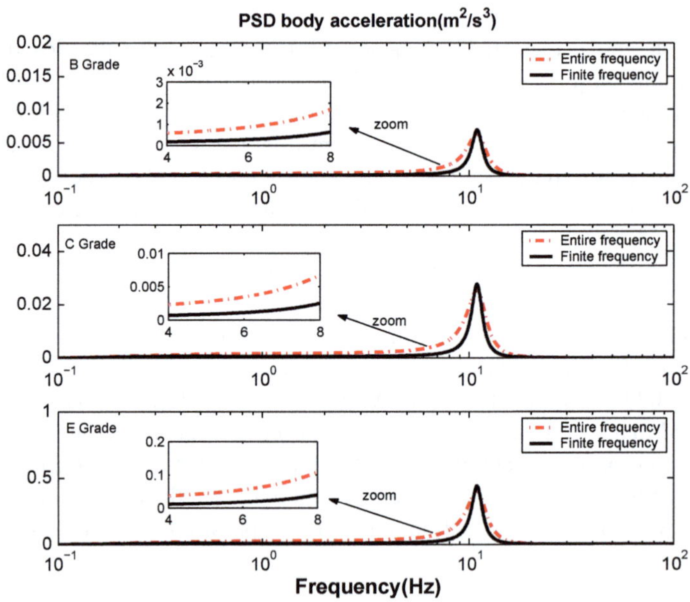

Fig. 6.8 The power spectral density of body acceleration with different road profiles ($d = 5$ ms)

the generalized Kalman–Yakubovich–Popov lemma. At the same time, the time-domain constraints have also been guaranteed in the controller design. Simulation results show that the finite frequency controllers achieve better disturbance attenuation performance over the concerned frequency range than those designed in the entire frequency.

6.2 Robust Sampled-Data H_∞ Control for Vehicle Active Suspension Systems

This section investigates the problem of robust sampled-data H_∞ control for active vehicle suspension systems. By using an input delay approach, the active vehicle suspension system with sampling measurements is transformed into a continuous-time system with a delay in the state. The transformed system contains non-differentiable time-varying state delay and polytopic parameter uncertainties. A Lyapunov functional approach is employed to establish the H_∞ performance, and the controller design is cast into a convex optimization problem with linear matrix inequality (LMI) constraints. A quarter-car model is also considered in the section and the effectiveness of the proposed approach is illustrated by a realistic design example.

6.2.1 Problem Formulation

In this section, The quarter car model shown in Fig. 1.1 is considered. The more details are provided in Sect. 1.2. Also, the effect of actuator dynamics is neglected and the actuator is modelled as an ideal force generator. According to the model, the motion equation of a quarter-car suspension system may be established as (1.5). Define the following state variables:

$$x_1(t) = z_s(t) - z_u(t),\ x_2(t) = z_u(t) - z_r(t),\ x_3(t) = \dot{z}_s(t),\ x_4(t) = \dot{z}_u(t), \tag{6.39}$$

where $x_1(t)$ denotes the suspension deflection, $x_2(t)$ is the tire deflection, $x_3(t)$ is the sprung mass speed, and $x_4(t)$ denotes the unsprung mass speed. We define disturbance input $w(t) = \dot{z}_r(t)$. Then, by defining

$$x(t) = \left[x_1(t)\ x_2(t)\ x_3(t)\ x_4(t) \right]^T,$$

the dynamic equations can be rewritten in the following state-space form:

$$\dot{x}(t) = A(\lambda)x(t) + B_1(\lambda)w(t) + B(\lambda)u(t), \tag{6.40}$$

where

$$A(\lambda)=\begin{bmatrix}0 & 0 & 1 & -1\\ 0 & 0 & 0 & 1\\ -\frac{k_s}{m_s} & 0 & -\frac{c_s}{m_s} & \frac{c_s}{m_s}\\ \frac{k_s}{m_u} & -\frac{k_t}{m_u} & \frac{c_s}{m_u} & -\frac{c_s+c_t}{m_u}\end{bmatrix},\ B(\lambda)=\begin{bmatrix}0\\ 0\\ \frac{1}{m_s}\\ -\frac{1}{m_u}\end{bmatrix},\ B_1(\lambda)=\begin{bmatrix}0\\ -1\\ 0\\ \frac{c_t}{m_u}\end{bmatrix},$$

where λ is used to characterize the parameter uncertainty, which will be described in detail subsequently.

According to the suspension performance requirements, we choose the H_∞ norm as performance measure and the body acceleration $\ddot{z}_s(t)$ as performance output, and choose the suspension stroke $z_s(t)-z_u(t)$ and relative dynamic tire load $k_t(z_u(t)-z_r(t))/(m_s+m_u)g$ as constrained outputs.

Therefore, the vehicle suspension control system can be described as:

$$\begin{aligned}\dot{x}(t) &= A(\lambda)x(t)+B_1(\lambda)w(t)+B(\lambda)u(t),\\ z_1(t) &= C_1(\lambda)x(t)+D_1(\lambda)u(t),\\ z_2(t) &= C_2(\lambda)x(t),\end{aligned}\tag{6.41}$$

where $A(\lambda)$, $B_1(\lambda)$ and $B(\lambda)$ are defined in (7.28), and

$$C_1(\lambda)=\begin{bmatrix}-\frac{k_s}{m_s} & 0 & -\frac{c_s}{m_s} & \frac{c_s}{m_s}\end{bmatrix},\ D_1(\lambda)=\frac{1}{m_s},\ C_2(\lambda)=\begin{bmatrix}1 & 0 & 0 & 0\\ 0 & \frac{k_t}{(m_s+m_u)g} & 0 & 0\end{bmatrix}.$$

By considering the modelling uncertainty, in this paper we assume the matrices $A(\lambda)$, $B(\lambda)$, $B_1(\lambda)$, $C_1(\lambda)$, $D_1(\lambda)$ and $C_2(\lambda)$ in (6.41) contain uncertain parameters, represented by λ. It is assumed that λ varies in a polytope of vertices $\lambda_1, \lambda_2, \ldots, \lambda_r$, i.e., $\lambda \in \Theta := Co\{\lambda_1, \lambda_2, \ldots, \lambda_r\}$, where the symbol Co denotes the convex hull, and thus we have

$$\Omega \triangleq (A(\lambda), B(\lambda), B_1(\lambda),\ C_1(\lambda), C_2(\lambda), D_1(\lambda)) \in \Theta,\tag{6.42}$$

where Θ is a given convex bounded polyhedral domain described by r vertices:

$$\Theta \triangleq \left\{\Omega \,\middle|\, \Omega=\sum_{i=1}^{r}\lambda_i\Omega_i;\quad \sum_{i=1}^{r}\lambda_i=1, \lambda_i\geq 0\right\},\tag{6.43}$$

with $\Omega_i \triangleq (A_i,\ B_i,\ B_{1i},\ C_{1i},\ C_{2i},\ D_{1i})$ denoting the vertices of the polytope.

It is assumed that the state variables of the active suspension system are measured at time instants $t_k, \ldots, t_{k+1}, \ldots$, that is, only $x(t_k)$ are available for interval $t_k \leq t < t_{k+1}$. We are interested in designing a state feedback controller of the form:

$$u(t)=u(t_k)=Kx(t_k),\qquad t_k\leq t<t_{k+1},\tag{6.44}$$

where K is the state feedback gain matrix to be designed.

Therefore, the closed-loop system is given by

$$\begin{aligned}\dot{x}(t) &= A(\lambda)x(t) + B_1(\lambda)w(t) + B(\lambda)Kx(t_k),\\ z_1(t) &= C_1(\lambda)x(t) + D_1(\lambda)Kx(t_k),\\ z_2(t) &= C_2(\lambda)x(t),\ t_k \le t < t_{k+1}.\end{aligned} \tag{6.45}$$

It is assumed that $w \in L_2[0,\infty)$, and without loss of generality, we have $\|w\|_2^2 \le w_{\max} < \infty$. Then, the objective of this section is to determine a controller gain K such that

(1) the closed-loop system is asymptotically stable;

(2) under zero initial condition, the closed-loop system guarantees that $\|z_1\|_2 < \gamma\|w\|_2$ for all nonzero $w \in L_2[0,\infty)$, where $\gamma > 0$ is a prescribed scalar;

(3) the following control output and input constraints are guaranteed:

$$\left|\{z_2(t)\}_j\right| \le \{z_{2,\max}\}_j,\ j = 1, 2,\ \ |u(t)| \le u_{\max},\ t > 0, \tag{6.46}$$

where $z_{2,\max} = \left[\ z_{\max}\ \right]^T$.

Before proceeding further, we first introduce the following general assumption.

Assumption 1 It is assumed that the interval between any two sampling instants is bounded by h ($h > 0$). That is, $t_{k+1} - t_k \le h$, $\forall k \ge 0$.

Remark 6.5 The closed-loop system in (6.45) involves both continuous and discrete signals. Due to parameter uncertainties, it is difficult to use the traditional lifting techniques to solve this sampled-data control problem.

Remark 6.6 It is known that ride comfort is frequency sensitive. Although our paper considers the ride comfort in full frequency, it is worth mentioning that our approach can be further extended to finite frequency case, by incorporating some frequency weighting transfer function.

6.2.2 *Constrained Sampled-Data Controller Design*

In this section, the problem formulated above will be solved by an input delay approach. The key idea behind this approach is that we represent the sampling instant t_k as

$$t_k = t - (t - t_k) = t - d(t),$$

where $d(t) = t - t_k$. Then, we obtain

$$u(t) = u(t_k) = u(t - d(t)),\ t_k \le t < t_{k+1}, \tag{6.47}$$

where $u(t_k)$ is a discrete-time control signal and the time-varying delay $d(t) = t - t_k \le h$ is piecewise-linear with derivative $\dot{d}(t) = 1$ for $t \ne t_k$. Recently, the H_∞

control problem for active vehicle suspension systems with input delay has been addressed in [4], where the time-delay is fixed and constant. It is worth pointing out that the transformed system in our problem contains non-differentiable time-varying delay in the states, which hinders the results in [4] to be directly applied to the problem considered here.

By making use of (6.47), the sampled-data formulation in (6.45) can be transformed into the following system:

$$\begin{aligned}\dot{x}(t) &= A(\lambda)x(t) + B(\lambda)Kx(t-d(t)) + B_1(\lambda)w(t),\\ z_1(t) &= C_1(\lambda)x(t) + D_1(\lambda)Kx(t-d(t)),\\ z_2(t) &= C_2(\lambda)x(t).\end{aligned} \tag{6.48}$$

Now, a continuous-time system with a time-varying delay $d(t)$ in the state, as shown in (6.48), has been obtained by transforming the sampled-data closed-loop system in (6.45) as above. In the following, we will investigate how to design a desired sampled-data controller based on the transformed closed-loop system in (6.48).

Theorem 6.7 *Consider the active suspension system in (6.45) under Assumption 1. Given scalars $\gamma > 0$, $h > 0$ and $\rho > 0$, if there exist matrices $P = P^T > 0$, $Q = Q^T > 0$ and S_i satisfying*

$$\begin{bmatrix} \Psi_{1i} + \Psi_{2i} + \Psi_{2i}^T + \Psi_{5i} & \sqrt{h}\Phi_{1i}^T & \sqrt{h}S_i & \Phi_{2i}^T \\ * & -Q^{-1} & 0 & 0 \\ * & * & -Q & 0 \\ * & * & * & -I \end{bmatrix} < 0, \; i = 1, \ldots, r, \tag{6.49}$$

$$\begin{bmatrix} -I & \sqrt{\rho}\{C_{2i}\}_j \\ * & -\{z_{2,\max}\}_j^2 P \end{bmatrix} < 0, \; i = 1, \ldots, r, \; j = 1, 2, \tag{6.50}$$

$$\begin{bmatrix} -I & \sqrt{\rho}K \\ * & -u_{\max}^2 P \end{bmatrix} < 0, \tag{6.51}$$

where

$$\begin{aligned}\Psi_{1i} &= \begin{bmatrix} PA_i + A_i^T P & PB_iK & PB_{1i} \\ * & 0 & 0 \\ * & * & 0 \end{bmatrix},\\ \Psi_{2i} &= \begin{bmatrix} S_i & -S_i & 0 \end{bmatrix}, \; \Psi_{5i} = \mathrm{diag}\begin{bmatrix} 0 & 0 & -\gamma^2 I \end{bmatrix},\\ \Phi_{1i} &= \begin{bmatrix} A_i & B_iK & B_{1i} \end{bmatrix}, \; \Phi_{2i} = \begin{bmatrix} C_{1i} & D_{1i}K & 0 \end{bmatrix},\end{aligned} \tag{6.52}$$

then a stabilizing controller in the form of (6.44) exists, such that

(1) the closed-loop system is asymptotically stable;

(2) under zero initial condition, the closed-loop system guarantees that $\|z_1\|_2 < \gamma\|w\|_2$ for all nonzero $w \in L_2[0, \infty)$;

(3) the control output and input constraints (6.46) are guaranteed with the disturbance energy under the bound $w_{\max} = (\rho - V(0))/\gamma^2$.

Proof Under the condition of the theorem, firstly, we show the asymptotic stability of (6.48) with $w(t) = 0$, that is

$$\dot{x}(t) = A(\lambda)x(t) + B(\lambda)Kx(t - d(t)). \tag{6.53}$$

Now, choose a Lyapunov functional candidate for system (6.53) as

$$V(t) = x^T(t)Px(t) + \int_{-h}^{0}\int_{t+\beta}^{t} \dot{x}^T(\alpha)Q\dot{x}(\alpha)d\alpha d\beta, \tag{6.54}$$

where $P > 0$ and $Q > 0$ are matrices to be determined.

The derivative of $V(t)$ satisfies

$$\dot{V}(t) = 2x^T(t)P\dot{x}(t) + h\dot{x}^T(t)Q\dot{x}(t) - \int_{t-h}^{t} \dot{x}^T(\alpha)Q\dot{x}(\alpha)d\alpha.$$

In addition, by the Newton–Leibniz formula, for any appropriately dimensioned matrices $\hat{S}(\lambda) = \sum_{i=1}^{r} \lambda_i \hat{S}_i = \sum_{i=1}^{r} \lambda_i \left[\, S_{1i}^T \; S_{2i}^T \,\right]^T$, we have

$$\zeta^T(t)\hat{S}(\lambda)\left[x(t) - x(t - d(t)) - \int_{t-d(t)}^{t} \dot{x}(\alpha)d\alpha\right] = 0,$$

where $\zeta(t) = \left[\, x^T(t) \; x^T(t - d(t)) \,\right]^T$. So we can get

$$\begin{aligned}\dot{V}(t) \le\; & 2x^T(t)P[A(\lambda)x(t) + B(\lambda)Kx(t - d(t))] - \int_{t-d(t)}^{t} \dot{x}^T(\alpha)Q\dot{x}(\alpha)d\alpha \\ & + h[A(\lambda)x(t) + B(\lambda)Kx(t - d(t))]^T Q[A(\lambda)x(t) + B(\lambda)Kx(t - d(t))] \\ & + 2\zeta^T(t)\hat{S}(\lambda)\left[x(t) - x(t - d(t)) - \int_{t-d(t)}^{t} \dot{x}(\alpha)d\alpha\right].\end{aligned} \tag{6.55}$$

Then, the time derivative of $V(t)$ along the solution of system (6.53) holds

$$\begin{aligned}\dot{V}(t) \le\; & \zeta^T(t)[E_1(\lambda) + E_2(\lambda) + E_2^T(\lambda) + E_3(\lambda) + h\hat{S}(\lambda)Q^{-1}\hat{S}^T(\lambda)]\zeta(t) \\ & - \int_{t-d(t)}^{t} [\zeta^T(t)\hat{S}(\lambda) + \dot{x}^T(\alpha)Q]Q^{-1}[\hat{S}^T(\lambda)\zeta(t) + Q\dot{x}(\alpha)]d\alpha,\end{aligned} \tag{6.56}$$

where

$$E_1(\lambda) = \begin{bmatrix} PA(\lambda) + A^T(\lambda)P & PB(\lambda)K \\ * & 0 \end{bmatrix},$$

$$E_2(\lambda) = \begin{bmatrix} \hat{S}(\lambda) & -\hat{S}(\lambda) \end{bmatrix}, \; E_3(\lambda) = h\begin{bmatrix} A(\lambda) & B(\lambda)K \end{bmatrix}^T Q \begin{bmatrix} A(\lambda) & B(\lambda)K \end{bmatrix}.$$

Note that $Q > 0$, thus $\int_{t-d(t)}^{t}[\zeta^T(t)\hat{S}(\lambda) + \dot{x}^T(\alpha)Q]Q^{-1}[\hat{S}^T(\lambda)\zeta(t) + Q\dot{x}(\alpha)]d\alpha$ is positive. By Schur complement, inequality (4.71) guarantees

$$E_{1i} + E_{2i} + E_{2i}^T + E_{3i} + h\hat{S}_i Q_i^{-1}\hat{S}_i^T < 0, \tag{6.57}$$

where

$$E_{1i} = \begin{bmatrix} PA_i + A_i^T P & PB_i K \\ * & 0 \end{bmatrix}, \; E_{2i} = \begin{bmatrix} \hat{S}_i & -\hat{S}_i \end{bmatrix}, \; E_{3i} = h\begin{bmatrix} A_i & B_i K \end{bmatrix}^T Q \begin{bmatrix} A_i & B_i K \end{bmatrix}.$$

According to the inner property of polytopic uncertain systems, and considering the form $A(\lambda) = \sum_{i=1}^{r} \lambda_i A_i$, $B(\lambda) = \sum_{i=1}^{r} \lambda_i B_i$, $\hat{S}(\lambda) = \sum_{i=1}^{r} \lambda_i \hat{S}_i$, from (6.57) we obtain

$$E_1(\lambda) + E_2(\lambda) + E_2^T(\lambda) + E_3(\lambda) + h\hat{S}(\lambda)Q^{-1}\hat{S}^T(\lambda) < 0. \tag{6.58}$$

Therefore, we have $\dot{V}(t) < 0$, and the asymptotic stability is established.

Next, we shall establish the H_∞ performance of the system in (6.48) under zero initial conditions. Firstly, define the Lyapunov functional as in (6.54). Then, by following similar lines as in the above proof, the time derivative of $V(t)$ is given by:

$$\begin{aligned} \dot{V}(t) \le\; & \bar{\zeta}^T(t)[\Psi_1(\lambda) + \Psi_2(\lambda) + \Psi_2^T(\lambda) + \Psi_3(\lambda) + hS(\lambda)Q^{-1}S^T(\lambda)]\bar{\zeta}(t) \\ & - \int_{t-d(t)}^{t} [\bar{\zeta}^T(t)S(\lambda) + \dot{x}^T(\alpha)Q]Q^{-1}[S^T(\lambda)\bar{\zeta}(t) + Q\dot{x}(\alpha)]d\alpha, \end{aligned} \tag{6.59}$$

where

$$\Psi_1(\lambda) = \begin{bmatrix} PA(\lambda) + A^T(\lambda)P & PB(\lambda)K & PB_1(\lambda) \\ * & 0 & 0 \\ * & * & 0 \end{bmatrix}, \; \Psi_2(\lambda) = \begin{bmatrix} S(\lambda) & -S(\lambda) & 0 \end{bmatrix},$$

$$\Psi_3(\lambda) = h\begin{bmatrix} A(\lambda) & B(\lambda)K & B_1(\lambda) \end{bmatrix}^T Q \begin{bmatrix} A(\lambda) & B(\lambda)K & B_1(\lambda) \end{bmatrix},$$

$$\bar{\zeta}(t) = \begin{bmatrix} x^T(t) & x^T(t-d(t)) & w^T(t) \end{bmatrix}^T, \; S(\lambda) = [\, S_1(\lambda) \; S_2(\lambda) \; S_3(\lambda) \,].$$

Thus, we have

$$\begin{aligned} & \dot{V}(t) + z_1^T(t)z_1(t) - \gamma^2 w^T(t)w(t) \\ \le\; & \bar{\zeta}^T(t)[\Psi_1(\lambda) + \Psi_2(\lambda) + \Psi_2^T(\lambda) + \Psi_3(\lambda) + hS(\lambda)Q^{-1}S^T(\lambda) + \Psi_4(\lambda) + \Psi_5(\lambda)]\bar{\zeta}(t) \\ & - \int_{t-d(t)}^{t} [\bar{\zeta}^T(t)S(\lambda) + \dot{x}^T(\alpha)Q]Q^{-1}[S^T(\lambda)\bar{\zeta}(t) + Q\dot{x}(\alpha)]d\alpha, \end{aligned} \tag{6.60}$$

where $\Psi_4(\lambda) = \begin{bmatrix} C_1(\lambda) & D_1(\lambda)K & 0 \end{bmatrix}^T \begin{bmatrix} C_1(\lambda) & D_1(\lambda)K & 0 \end{bmatrix}$, $\Psi_5(\lambda) = \text{diag}\{0,\ 0,\ -\gamma^2 I\}$. By Schur complement and inner property of polytopic uncertain systems, (6.49) together with (6.43) guarantees

$$\Psi_1(\lambda) + \Psi_2(\lambda) + \Psi_2^T(\lambda) + \Psi_3(\lambda) + hS(\lambda)Q^{-1}S^T(\lambda) + \Psi_4(\lambda) + \Psi_5(\lambda) < 0. \tag{6.61}$$

Thus, from (6.60) we get

$$\dot{V}(t) + z_1^T(t)z_1(t) - \gamma^2 w^T(t)w(t) < 0, \tag{6.62}$$

for all nonzero $w \in L_2[0, \infty)$. Under zero initial conditions, we have $V(0) = 0$ and $V(\infty) \geq 0$. Integrating both sides of (6.62) yields $\|z_1\|_2 < \gamma \|w\|_2$ for all nonzero $w \in L_2[0, \infty)$, and the H_∞ performance is established.

In the following, we will show that the hard constraints are guaranteed. Inequality (6.62) guarantees $\dot{V}(t) - \gamma^2 w^T(t)w(t) < 0$. Integrating both sides of the above inequality from zero to any $t > 0$, we obtain

$$V(t) - V(0) < \gamma^2 \int_0^t w^T(t)w(t)dt < \gamma^2 \|w\|_2^2. \tag{6.63}$$

Noting that the second term of the Lyapunov functional (6.54) is positive, we obtain $x^T(t)Px(t) < \rho$, with $\rho = \gamma^2 w_{\max} + V(0)$. It is also true that: $x^T(t - d(t))Px(t - d(t)) < \rho$, with $t > d(t)$. Consider

$$\begin{aligned} \max_{t>0} \left|\{z_2(t)\}_j\right|^2 &= \max_{t>0} \left\|x^T(t)\{C_{2i}\}_j^T\{C_{2i}\}_j x(t)\right\|_2 \\ &= \max_{t>0} \left\|x^T(t)P^{\frac{1}{2}}P^{-\frac{1}{2}}\{C_{2i}\}_j^T\{C_{2i}\}_j P^{-\frac{1}{2}}P^{\frac{1}{2}}x(t)\right\|_2 \\ &< \rho \cdot \theta_{\max}(P^{-\frac{1}{2}}\{C_{2i}\}_j^T\{C_{2i}\}_j P^{-\frac{1}{2}}), \\ & i = 1, \ldots, r,\ \ j = 1, 2, \end{aligned} \tag{6.64}$$

$$\begin{aligned} \max_{t>0} |u(t)|^2 &= \max_{t_k>0} |u(t_k)|^2 = \max_{t>d(t)} \left\|x^T(t - d(t))K^T K x(t - d(t))\right\|_2 \\ &< \rho \cdot \theta_{\max}(P^{-\frac{1}{2}}K^T K P^{-\frac{1}{2}}), \end{aligned} \tag{6.65}$$

where $\theta_{\max}(\cdot)$ represents maximal eigenvalue. From the above inequalities, we know that the constraints (6.46) are guaranteed, if

$$\rho \cdot P^{-\frac{1}{2}}\{C_{2i}\}_j^T\{C_{2i}\}_j P^{-\frac{1}{2}} < \{z_{2,\max}\}_j^2 I,\ i = 1, \ldots, r,\ j = 1, 2, \tag{6.66}$$

$$\rho \cdot P^{-\frac{1}{2}}K^T K P^{-\frac{1}{2}} < u_{\max}^2 I. \tag{6.67}$$

By Schur complement, (6.66) and (6.67) are equivalent to (6.50) and (6.51), and the proof is completed. □

Define $J_1 = \text{diag}\left\{P^{-1},\ P^{-1},\ I,\ I,\ P^{-1},\ I\right\}$, $J_2 = \text{diag}\left\{I,\ P^{-1}\right\}$, $J_3 = \text{diag}\left\{P^{-1},\ P^{-1},\ I\right\}$. Pre- and post-multiplying (6.49)–(6.51) by J_1^T, J_2^T, J_2^T and their transposes, respectively, together with the change of matrix variables defined by $\bar{P} = P^{-1}$, $\bar{K} = KP^{-1}$, $\bar{Q} = P^{-1}QP^{-1}$, $\bar{S}_i = J_3 S_i \bar{P}$, we obtain the following result by noting that $-\bar{P}\bar{Q}^{-1}\bar{P} \leq \bar{Q} - 2\bar{P}$.

Theorem 6.8 *Consider the active suspension system in (6.45) under Assumption 1. Given scalars $\gamma > 0$, $h > 0$ and $\rho > 0$, if there exist symmetric matrices $\bar{P} > 0$, $\bar{Q} > 0$, and matrices $\bar{S}_i$, $\bar{K}$ satisfying*

$$\begin{bmatrix} \bar{\Psi}_{1i} + \bar{\Psi}_{2i} + \bar{\Psi}_{2i}^T + \Psi_{5i} & \sqrt{h}\bar{\Phi}_{1i}^T & \sqrt{h}\bar{S}_i & \bar{\Phi}_{2i}^T \\ * & \bar{Q} - 2\bar{P} & 0 & 0 \\ * & * & -\bar{Q} & 0 \\ * & * & * & -I \end{bmatrix} < 0,\ i = 1, \ldots, r, \tag{6.68}$$

$$\begin{bmatrix} -I & \sqrt{\rho}\{C_{2i}\}_j \bar{P} \\ * & -\{z_{2,\max}\}_j^2 \bar{P} \end{bmatrix} < 0,\ i = 1, \ldots, r,\ j = 1, 2, \tag{6.69}$$

$$\begin{bmatrix} -I & \sqrt{\rho}\bar{K} \\ * & -u_{\max}^2 \bar{P} \end{bmatrix} < 0, \tag{6.70}$$

where

$$\begin{aligned} \bar{\Psi}_{1i} &= \begin{bmatrix} A_i\bar{P} + \bar{P}A_i^T & B_i\bar{K} & B_{1i} \\ * & 0 & 0 \\ * & * & 0 \end{bmatrix}, \\ \bar{\Psi}_{2i} &= \begin{bmatrix} \bar{S}_i & -\bar{S}_i & 0 \end{bmatrix},\ \Psi_{5i} = \text{diag}\begin{bmatrix} 0 & 0 & -\gamma^2 I \end{bmatrix}, \\ \bar{\Phi}_{1i} &= \begin{bmatrix} A_i\bar{P} & B_i\bar{K} & B_{1i} \end{bmatrix},\ \bar{\Phi}_{2i} = \begin{bmatrix} C_{1i}\bar{P} & D_{1i}\bar{K} & 0 \end{bmatrix}, \end{aligned} \tag{6.71}$$

then a stabilizing controller in the form of (6.44) exists, such that

(1) the closed-loop system is asymptotically stable;

(2) under zero initial condition, the closed-loop system guarantees that $\|z_1\|_2 < \gamma \|w\|_2$ for all nonzero $w \in L_2[0, \infty)$;

(3) the control output and input constraints (6.46) are guaranteed with the disturbance energy under the bound $w_{\max} = (\rho - V(0))/\gamma^2$.

Moreover, if inequalities (6.68)–(6.70) have a feasible solution, then the control gain K in (6.44) is given by

$$K = \bar{K}\bar{P}^{-1}. \tag{6.72}$$

Remark 6.9 The conditions in Theorem 6.8 are LMIs not only over the matrix variables, but also over the scalar γ. This implies that the scalar γ can be included as an optimization variable to obtain a reduction of the guaranteed H_∞ performance bound. Then the minimal γ can be found by solving the following convex optimiza-

Table 6.1 Quarter-Car model parameters

m_s	m_u	k_s	k_t	c_s	c_t
973 kg	114 kg	42720 N/m	101115 N/m	1095 Ns/m	14.6 Ns/m

tion problem: minimize γ subject to (6.68)–(6.70) over $\bar{P} > 0$, $\bar{Q} > 0$, $\bar{S}_i > 0$ and $\bar{K}$.

6.2.3 Simulation Verification

In this section, we provide an example to illustrate the effectiveness of the proposed sampled-data H_∞ controller design method. The quarter-car model parameters are borrowed from [4] and listed in Table 6.1.

Firstly, we consider the nominal system, whose parameter matrices have no uncertainties. Assume the maximum allowable suspension stroke $z_{\max} = 0.08$ m, the maximum force output $u_{\max} = 1500$ N, the sampling interval $h = 10$ ms. Here, we choose $\rho = 1$ (the detailed discussion of its selection is given in [5]). By solving the convex optimization problem formulated in the above section, the minimum guaranteed closed-loop H_∞ performance obtained is $\gamma_{\min} = 8.6758$. Then, an admissible control gain matrix is $K = 10^3 \times \begin{bmatrix} 0.7646 & 3.6362 & -5.3292 & -0.5438 \end{bmatrix}$.

In the following, we will illustrate the performance of the closed-loop sampled-data suspension system. Evaluation of the vehicle suspension performance is based on the examination of three response quantities, that is, the sprung mass acceleration $z_1(t)$, the suspension deflection and the tire deflection, which can be shown from $z_2(t)$. A controller is to be designed such that: (1) the sprung mass acceleration $z_1(t)$ is as small as possible; (2) the suspension deflection is below the maximum allowable suspension stroke $z_{\max} = 0.08$ m; (3) the controlled output defined satisfy $\{z_2(t)\}_2 < 1$; and (4) the active force $|u(t)| \leq u_{\max}$. In order to evaluate the suspension characteristics with respect to ride comfort, vehicle handling, and working space of the suspension, the variability of the road profiles is taken into account. In the context of vehicle suspension performance, road disturbances can be generally assumed as shocks. Shocks are discrete events of relatively short duration and high intensity, caused by, for example, a pronounced bump or pothole on an otherwise smooth road. In the following, a kind of road profile is used to validate the performance of the presented control approach.

Now consider the case of an isolated bump in an otherwise smooth road surface in [6]. The corresponding ground displacement is given by

$$z_r(t) = \begin{cases} \frac{A}{2}(1 - \cos(\frac{2\pi V}{L}t)), & \text{if } 0 \leq t \leq \frac{L}{V}, \\ 0, & \text{if } t > \frac{L}{V}, \end{cases} \tag{6.73}$$

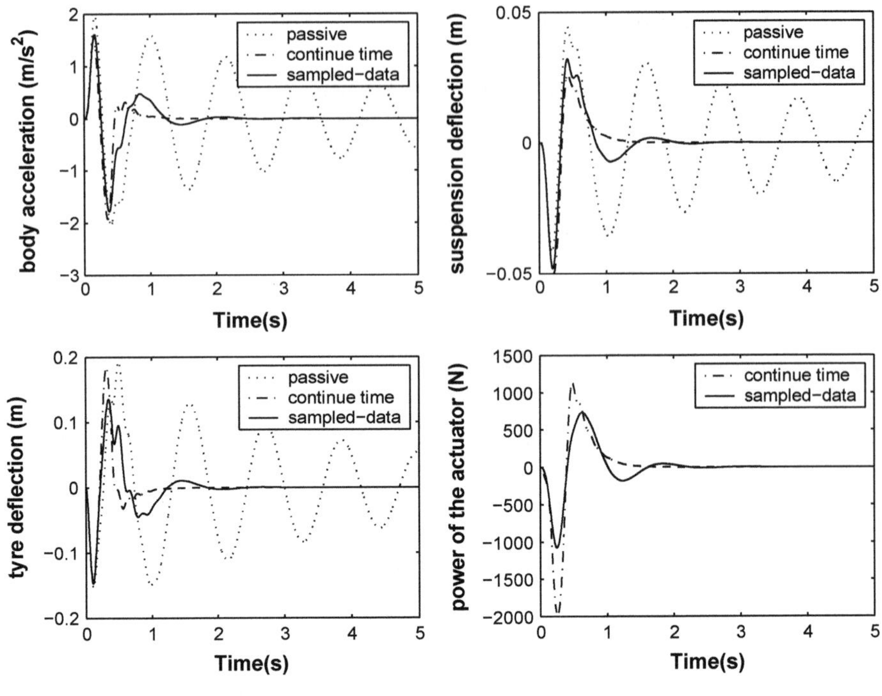

Fig. 6.9 Bump response for nominal system

where A and L are the height and the length of the bump. Assume $A = 60\,\text{mm}$, $L = 5\,\text{m}$ and the vehicle forward velocity as $V = 45\,(\text{km/h})$, which implies a disturbance energy of $0.0444\ \text{m}^2/\text{s}$. In order to compare, we give the controller of continue-time system which is obtained from [4]:

$$K_c = 10^4 \times \begin{bmatrix} -8.9220 & -0.1447 & -3.6650 & 0.1491 \end{bmatrix}.$$

The responses of the open-loop system ($u(t_k) = 0$, passive mode), the continue-time closed-loop system ($u(t) = K_c x(t)$) and the sampled-data closed-loop system (active mode) which is composed by the controller we designed above are depicted in Fig. 6.9, which shows the bump response of the body acceleration, the suspension deflection, the tire deflection, and the active force, respectively, with the sampling interval $h = 10$ ms, where the passive suspension, the continue-time active suspension and sampled-data active suspension are depicted in point lines, point-solid lines and solid lines, respectively. From these figures, we can see that the sampled-data controller yields the least value of the maximum body acceleration, compared with the open-loop system and the continuous-time controller. In addition, we can see that the active control force constraint is respected by the sampled-date control, while not respected by the continuous-time controller due to its ignorance of the hard constraints in the controller design process.

Table 6.2 Guaranteed H_∞ performances for different sampling intervals

h(ms)	5	10	15	20	25
$\gamma_{\min}$	8.3651	8.6758	9.7879	12.4934	21.6790

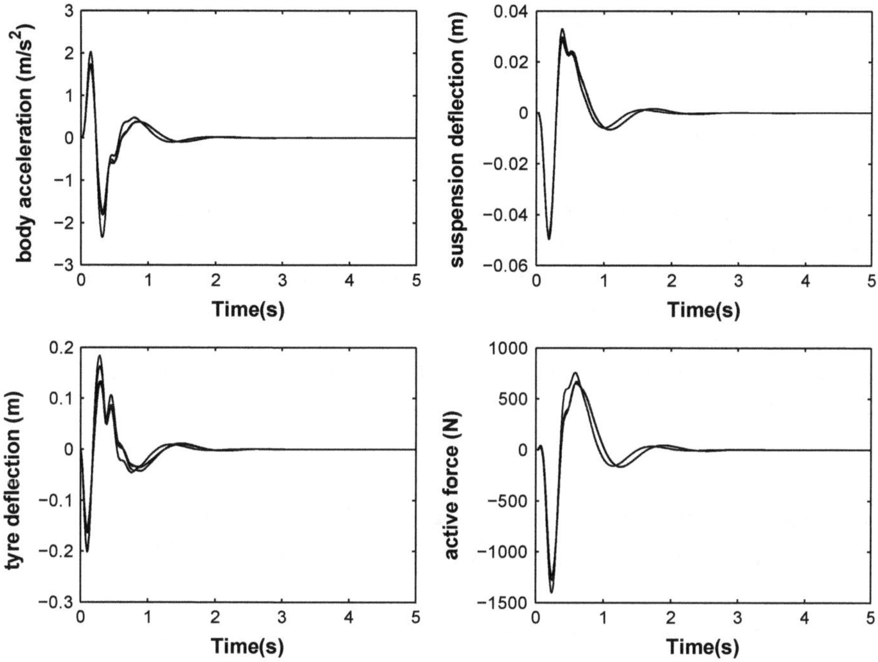

Fig. 6.10 Bump response with different λ

It is interesting to note that the sampling period has much to do with the guaranteed performance $\gamma_{\min}$ we obtain by the convex optimization problem formulated in the above section. Table 6.2 lists the guaranteed performance $\gamma_{\min}$ we obtain for different sampling periods, from which we can see that the guaranteed performance is larger when the sampling period increases.

Now, we consider the uncertain case, that is, we design robust sampled-data controllers for uncertain suspension systems. Assume that the sprung mass m_s and the unsprung m_u contain uncertainties, which are expressed as $m_s = (973 + \lambda_1)$ kg, $m_u = (114 + \lambda_2)$ kg, where λ_1 and λ_2 satisfy $|\lambda_1| \leq \bar{\lambda}_1$ and $|\lambda_2| \leq \bar{\lambda}_2$. It is assumed that $\bar{\lambda}_1 = 100$ and $\bar{\lambda}_2 = 10$. In this case, the suspension system can be represented by a four-vertex polytopic system. By solving the corresponding convex optimization problem, when the sampling period $h = 10$ ms, the minimum guaranteed closed-loop H_∞ performance obtained (in terms of the feasibility of (6.68)–(6.70)) is $\gamma_{\min} = 10.4220$. An admissible robust control gain matrix is given by $K = 10^3 \times \left[\, 1.3308 \; 3.9564 \; -5.5819 \; 0.5438 \,\right]$.

In the following, we will illustrate the performance of the closed-loop sampled-data suspension system with parameter uncertainties. Figure 6.10 depicts the bump response of the body acceleration, and the constrained conditions (suspension deflection, tire deflection and active force) with the sampling interval $h = 10$ ms. In each figure, the four vertex systems are depicted with the sampling interval $h = 10$ ms. The effectiveness of the control design is apparent from these figures.

6.2.4 Conclusion

In this section, the problem of robust sampled-data H_∞ control for uncertain active vehicle suspension systems has been investigated. By using an input delay approach, the active vehicle suspension system with sampling measurements has been transformed into a continuous-time system with a delay in the state, and polytopic parameter uncertainty has been utilized to characterize the real uncertain situation. A quarter-car model has been considered and the effectiveness of the proposed approach has been illustrated by a practical design example.

6.3 Reliability Control for Uncertain Half-Car Active Suspension Systems with Possible Actuator Faults

In this section, active suspension systems have received increased importance for improving automotive safety and comfort. In active suspensions, actuators are placed between the car body and wheel-axle, and are able to both add and dissipate energy from the system, which enables the suspension to control the attitude of the vehicle, to reduce the effects of the vibrations, and then to increase ride comfort and vehicle road handling. However, the attained benefits are paralleled with the increasing possibility of component failures. In this section, a fault tolerant control approach is proposed to deal with the problem of fault accommodation for unknown actuator failures of active suspension systems, where an adaptive robust controller is designed to adapt and compensate the parameter uncertainties, external disturbances and uncertain nonlinearities generated by the system itself and actuator failures. Comparative simulation studies are then given to illustrate the effectiveness of the proposed controllers.

6.3.1 Stability Analysis

In this section, the nonlinear half-car model is the second model in Sect. 4.2, where more details are provided. This model has been used extensively in the literature and captures many important characteristics of vertical and pitch motions. Here, the effect of lateral motion is neglected. The ideal dynamic equations of the sprung and unsprung masses are given by (4.43) in Chap. 4. The forces produced by the nonlinear stiffening spring, the piece-wise linear damper and the tire obey (4.44). The definitions of state variables and the state-space expression of the nonlinear half-car model is derived as (4.49) in Chap. 4.

In this study, we assume that system is subjected to actuator faults which can be broadly divided into two categories in [7, 8]:

1. failures that result in a total loss of effectiveness of the control actuator, including lock-in-place, float, and hard-over failure, which can be represented as $u_i = \bar{u}_i, \forall t \geq T_{fi}$, if actuator gets stuck at the unknown instant T_{fi}. u_i^* $(i = 1, 2)$ represents the final control command to the actuator, $\bar{u}$ is an unknown constant value at which the actuator gets lock-in-place, float, or hard-over failure.
2. failures that cause partial loss of effectiveness, which can be represented as $u_i = \eta_i u_i^*, \forall t \geq T_{fi}$, if actuator loses efficiency at T_{fi}. $\eta_i \in \left[\, \eta_{i\min} \;\; 1 \,\right]$ represents actuator loss in efficiency, where $\eta_{i\min}$ is a known constant standing for the minimum value of η_i.

With a fault model stated above, the lumped form of the control inputs can be given as

$$u_i = \eta_i(1 - \sigma_i)u_i^* + \sigma_i \bar{u}_i, \tag{6.74}$$

where $\sigma_i = 1$ corresponds to the total loss of effectiveness of the control actuator, $\sigma_i = 0$ and $\eta_{i\min} \leq \eta_i < 1$ represents actuator partial loss of efficiency, $\sigma_i = 0$ and $\eta_i = 1$ corresponds to the healthy actuator. With this, we can rewrite the system as

$$\begin{aligned}
M\ddot{z}_c &= \Psi_1(t) + \sum_{i=1}^{2}(\alpha_i u_i^* + \beta_i) + F_l,\\
I\ddot{\varphi} &= \Psi_2(t) + a(\alpha_1 u_1^* + \beta_1) - b(\alpha_2 u_2^* + \beta_2) + F_\varphi,\\
m_f \ddot{z}_1 &= F_{sf} + F_{df} - F_{tf} - F_{bf} - \alpha_1 u_1^* - \beta_1,\\
m_r \ddot{z}_2 &= F_{sr} + F_{dr} - F_{tr} - F_{br} - \alpha_2 u_2^* - \beta_2,
\end{aligned} \tag{6.75}$$

where $\alpha_i = \eta_i(1 - \sigma_i)$ and $\beta_i = \sigma_i \bar{u}_i$, $i = 1, 2$. Note that α_i is the unknown measure of actuator effectiveness after faults and β_i is the unknown measure of the fault magnitude which needs to be compensated.

In this section, the focus will be on a class of failure scenarios that satisfy the following realistic assumptions.

Assumption 2 Up to one actuator can undergo total failure, such as lock-in-place, float and hard-over failure.

Assumption 3 The extent of the uncertain parameter is known, i.e.,

$$
\begin{aligned}
M \in \Omega_M &= \{M : M_{\min} \leq M \leq M_{\max}\}, \\
I \in \Omega_I &= \{I : I_{\min} \leq I \leq I_{\max}\}, \\
\alpha_i \in \Omega_{\alpha_i} &= \{\alpha_i : \alpha_{i\min} \leq \alpha_i \leq \alpha_{i\max}\}, \\
\beta_i \in \Omega_{\beta_i} &= \{\beta_i : \beta_{i\min} \leq \beta_i \leq \beta_{i\max}\}.
\end{aligned}
$$

Assumption 4 The uncertain nonlinear friction forces F_l and F_φ satisfy

$$
\begin{aligned}
F_l \in \Omega_{F_l} &= \{F_l : |F_l| \leq \delta_1(t)\}, \\
F_\varphi \in \Omega_{F_\varphi} &= \left\{F_\varphi : \left|F_\varphi\right| \leq \delta_2(t)\right\},
\end{aligned}
$$

where $\delta_i(t)$, $i = 1, 2$ are unknown but bounded functions.

Assumption 5 The reference trajectories r_z and r_φ are smooth functions, the magnitudes of which are bounded.

For the active suspension systems subjected to actuator fault, our goal is to synthesize adaptive robust control laws to stabilize the vertical and pitch motions of the suspension systems in the presence of parametric uncertainties and uncertain nonlinearities.

6.3.2 Control Law Synthesis

The problem formulated previously will be solved by designing an adaptive robust controller with the following functions: (1) the proposed controller should bring the greatest degree of improvement in ride comfort, although there are uncertain parameters and external disturbances in controller design; (2) the actuator fault will be considered in the controller design such that the closed-loop system has the ability of fault tolerant. The whole process of controller design can be given as follows.

Adaptive Robust Fault Tolerant Controller Design

Define $p_i = \dot{e}_i + k_i e_i$, $i = z, \varphi$, where $e_z = z_c - r_z$ and $e_\varphi = \varphi - r_\varphi$ are tracking errors from the vertical and pitch displacements to their corresponding reference trajectories, and $k_i > 0$ represents tunable gains. It can be known that e_i will converge to zero or be bounded as long as p_i converges to zero or be bounded, because the transfer function $\frac{e_i(s)}{p_i(s)} = \frac{1}{s+k_i}$ is stable. Then, the control target in this step switches to keep p_i converge to zero or be bounded.

Before proceeding, motivated by [9], a projection-based adaptive law is introduced firstly, which will be used to tackle unknown parameters. For an uncertain parameter θ, following Assumption 6, the estimate parameter θ is updated through the following adaptive law:

$$
\dot{\hat{\theta}} = Proj_\theta(r_\theta \tau_\theta) = \begin{cases} 0, \ \text{if } \hat{\theta} = \theta_{\max}, \ r_\theta \tau_\theta > 0, \\ 0, \ \text{if } \hat{\theta} = \theta_{\min}, \ r_\theta \tau_\theta < 0, \\ r_\theta \tau_\theta, \ \text{otherwise}, \end{cases}
$$

which, for any adaptation function τ_θ, has the following two properties

$$\begin{aligned}
&\text{Property 1: } \tilde{\theta}(r_\theta^{-1} Proj_\theta(r_\theta \tau_\theta) - \tau_\theta) \le 0, \ \forall \tau_\theta, \\
&\text{Property 2: } \hat{\theta} \in \Omega_{\hat{\theta}} = \left\{\hat{\theta} : \hat{\theta}_{\min} \le \hat{\theta} \le \hat{\theta}_{\max}\right\}.
\end{aligned}$$

Firstly, we can obtain the differentiations of the dynamics $p_i = \dot{e}_i + k_i e_i$, $i = z, \varphi$, with the form as:

$$\begin{aligned}
M\dot{p}_z &= \Psi_1(t) + \sum_{i=1}^{2}(\alpha_i u_i^* + \beta_i) + F_l - M\ddot{r}_z + Mk_z\dot{e}_z \\
&= \Psi_1(t) + \sum_{i=1}^{2}(\hat{\alpha}_i u_i^* + \hat{\beta}_i) + F_l - \hat{M}(\ddot{r}_z - k_z\dot{e}_z) \\
&\quad - \sum_{i=1}^{2}(\tilde{\alpha}_i u_i^* + \tilde{\beta}_i) + \tilde{M}(\ddot{r}_z - k_z\dot{e}_z), \qquad (6.76)\\
I\dot{p}_\varphi &= \Psi_2(t) + a(\alpha_1 u_1^* + \beta_1) - b(\alpha_2 u_2^* + \beta_2) + F_\varphi \\
&\quad - I\ddot{r}_\varphi + Ik_\varphi\dot{e}_\varphi \\
&= \Psi_2(t) + a(\hat{\alpha}_1 u_1^* + \hat{\beta}_1) - b(\hat{\alpha}_2 u_2^* + \hat{\beta}_2) + F_\varphi \\
&\quad - \hat{I}(\ddot{r}_\varphi - k_\varphi\dot{e}_\varphi) - a(\tilde{\alpha}_1 u_1^* + \tilde{\beta}_1) \\
&\quad + b(\tilde{\alpha}_2 u_2^* + \tilde{\beta}_2) + \tilde{I}(\ddot{r}_\varphi - k_\varphi\dot{e}_\varphi). \qquad (6.77)
\end{aligned}$$

Design an adaptive robust control law as

$$u_i^* = u_{ia}^* + u_{is}^*, \qquad (6.78)$$

where u_{ia}^* is used to achieve an improved adaptive model compensation and u_{is}^* functions as the robust term. The adaptive part u_{ia}^* is designed as:

$$\begin{aligned}
u_{1a}^* &= \frac{1}{(a+b)\hat{\alpha}_1}\Big(-b\Psi_1(t) - (a+b)\hat{\beta}_1 + \hat{I}(\ddot{r}_\varphi - k_\varphi\dot{e}_\varphi) \\
&\quad - \Psi_2(t) + b\hat{M}(\ddot{r}_z - k_z\dot{e}_z) - bk_{pz}p_z - k_{p\varphi}p_\varphi\Big), \qquad (6.79)\\
u_{2a}^* &= \frac{1}{(a+b)\hat{\alpha}_2}\Big(-a\Psi_1(t) - (a+b)\hat{\beta}_2 - \hat{I}(\ddot{r}_\varphi - k_\varphi\dot{e}_\varphi) \\
&\quad + \Psi_2(t) + a\hat{M}(\ddot{r}_z - k_z\dot{e}_z) - ak_{pz}p_z + k_{p\varphi}p_\varphi\Big), \qquad (6.80)
\end{aligned}$$

where k_{pz} and $k_{p\varphi}$ are used for tuning the controller, and the adaptation laws are chosen as the projection type with the following form:

$$\dot{\hat{\alpha}}_i = Proj_{\alpha_i}(r_{\alpha_i}\tau_{\alpha_i}),\ \dot{\hat{\beta}}_i = Proj_{\hat{\beta}_i}(r_{\beta_i}\tau_{\beta_i}), \tag{6.81}$$

$$\dot{\hat{M}} = Proj_M(r_M\tau_M),\ \dot{\hat{I}} = Proj_I(r_I\tau_I), \tag{6.82}$$

where r_{α_i}, r_{β_i}, r_M, r_I are tunable adaptation gains and

$$\begin{aligned}\tau_{\alpha 1} &= (p_z + ap_\varphi)u_1^*,\ \tau_{\alpha 2} = (p_z - bp_\varphi)u_2^*,\\ \tau_{\beta_1} &= p_z + ap_\varphi,\ \tau_{\beta_2} = p_z - bp_\varphi,\\ \tau_M &= -p_z(\ddot{r}_z - k_z\dot{e}_z),\ \tau_I = -p_\varphi(\ddot{r}_\varphi - k_\varphi\dot{e}_\varphi).\end{aligned}$$

The robust control function u_{is}^* is now chosen to satisfy the following conditions:

$$\begin{aligned}&\text{condition 1: } \begin{matrix} p_z\{\sum_{i=1}^2 \hat{\alpha}_i u_{is}^* - \sum_{i=1}^2(\tilde{\alpha}_i u_i^* + \tilde{\beta}_i) + F_l \\ +\tilde{M}(\ddot{r}_z - k_z\dot{e}_z)\} \le \varepsilon_1,\end{matrix}\\ &\text{condition 2: } p_z \textstyle\sum_{i=1}^2 \hat{\alpha}_i u_{is}^* \le 0,\\ &\text{condition 3: } \begin{matrix} p_\varphi\{a\hat{\alpha}_1 u_{1s}^* - b\hat{\alpha}_2 u_{2s}^* - a(\tilde{\alpha}_1 u_1^* + \tilde{\beta}_1) + F_\varphi \\ +b(\tilde{\alpha}_2 u_2^* + \tilde{\beta}_2) + \tilde{I}(\ddot{r}_\varphi - k_\varphi\dot{e}_\varphi)\} \le \varepsilon_2,\end{matrix}\\ &\text{condition 4: } p_\varphi(a\hat{\alpha}_1 u_{1s}^* - b\hat{\alpha}_2 u_{2s}^*) \le 0,\end{aligned}$$

where ε_1 and ε_2 are designed parameters which can be arbitrarily small. Following [10], the robust control part u_{is}^* can be chosen as:

$$\begin{aligned}u_{1s}^* &= -\frac{bh_1(t)p_z}{4(a+b)\alpha_{1\min}} - \frac{h_2(t)p_\varphi}{4(a+b)\alpha_{1\min}},\\ u_{2s}^* &= -\frac{ah_1(t)p_z}{4(a+b)\alpha_{2\min}} + \frac{h_2(t)p_\varphi}{4(a+b)\alpha_{2\min}},\end{aligned}$$

where $h_i(t)$ be any smooth function satisfying

$$\begin{aligned}h_1(t) \ge\ & \frac{1}{\epsilon_{1a}}\Bigg\{\sum_{i=1}^2(\beta_{i\max} - \beta_{i\min})\\ & +(M_{\max} - M_{\min})\cdot|\ddot{r}_z - k_z\dot{e}_z|\\ & +\sum_{i=1}^2\left[(\alpha_{i\max} - \alpha_{i\min})\cdot\left|u_{ia}^*\right|\right]\Bigg\}^2 + \frac{1}{\epsilon_{1b}}\delta_1^2(t),\\ h_2(t) \ge\ & \frac{1}{\epsilon_{2a}}\left\{a\left[\alpha_{1\max} - \alpha_{1\min})\cdot\left|u_{1a}^*\right| + (\beta_{1\max} - \beta_{1\min})\right]\right.\\ & +b\left[\alpha_{2\max} - \alpha_{2\min})\cdot\left|u_{2a}^*\right| + (\beta_{2\max} - \beta_{2\min})\right]\\ & \left.+(I_{\max} - I_{\min})\cdot\left|\ddot{r}_\varphi - k_\varphi\dot{e}_\varphi\right|\right\}^2 + \frac{1}{\epsilon_{2b}}\delta_2^2(t),\end{aligned}$$

and ϵ_{ia} and ϵ_{ib} are adjustable small positive numbers satisfying $\epsilon_{ia} + \epsilon_{ib} = \epsilon_i$.

The stability of zero dynamics should be Guaranteed. The adaptive robust design yields to a 4th order error dynamic, while the original system is a 8th order system. So the zero dynamics consist of four states. It is easy to verify that the zero dynamics are stable, which can be seen in [11].

Remark 6.10 The reference trajectories are planned as follows. One can change the pre-determined time T_r to adjust the vertical acceleration of car body to high or low levels, and thus, contribute to ride comfort.

$$r_z(t) = \begin{cases} a_0 + a_1 t + a_2 t^2 + a_3 t^3 + a_4 t^4, & t < T_r \\ 0, & t \geq T_r \end{cases},$$

where coefficient vectors $a_i, i = 0, 1, 2, 3, 4$ are determined such that

$$\begin{aligned} r_z(0) &= a_0 = x_1(0), \\ \dot{r}_z(0) &= a_1 = x_2(0), \\ r_z(T_r) &= a_0 + a_1 T_r + a_2 T_r^2 + a_3 T_r^3 + a_4 T_r^4 = 0, \\ \dot{r}_z(T_r) &= a_1 + 2a_2 T_r + 3a_3 T_r^2 + 4a_4 T_r^3 = 0, \\ \ddot{r}_z(T_r) &= 2a_2 + 6a_3 T_r + 12a_4 T_r^2 = 0, \end{aligned} \tag{6.83}$$

which can guarantee that (1) $\dot{e}_z(0) = e_z(0) = 0$; (2) the vector function $r_z(t) \in C_3$. Furthermore, it is easy to see from (6.83) that $r_z(t) = 0$ and $\dot{r}_z(t) = 0$ can be reached in a pre-determined time T_r. As discussed previously, in this paper, one of the main targets is to control the vertical displacement of car body $z_s(t)$ as small as possible, and it is better to keep $z_s(t)$ converge to zero. In this case, the reference trajectory should be chosen as zero. However, zero reference trajectory may result in large stable time and large body acceleration response, which will also reduce the ride comfort. As a result, instead of zero, this paper employ a decreasing polynomial as the reference trajectory, and this polynomial contains a tuning parameter T_r which can adjust to large or small value to change the trajectory. Theoretically, T_r can be set as arbitrarily small value to make the vertical displacement vanish fast. However, smaller settle time T_r results in larger vertical accelerations and larger tracking error e_z, which can be confirmed in the following simulations. The reference trajectory of pitch angle is chosen as $r_\varphi(t) = 0$.

Main Results

Based on the previous design process, we can conclude the following theorem.

Theorem 6.11 *For the active suspension systems subjected to actuator fault, with the ARC law (6.78) and the projection type adaptive laws (6.81) and (6.82), the following results hold:*

A. In general (i.e., the system is subjected to parametric uncertainties, unmodelled uncertainties and external disturbances), the tracking errors p_z and p_φ are bounded;

B. If after a finite time, the system is subjected to parametric uncertainties only, then the tracking errors p_z and p_φ will converge to zero in a finite time.

Proof (*Proof of A*) Choose a positive definite function as

$$V = \frac{1}{2}Mp_z^2 + \frac{1}{2}Ip_\varphi^2,$$

and we have

$$\dot{V} = Mp_z\dot{p}_z + Ip_\varphi\dot{p}_\varphi. \tag{6.84}$$

Substituting the ARC law (6.78) into (6.84) results in

$$\dot{V}(t) \le -k_z p_z^2 - k_\varphi p_\varphi^2 + \varepsilon_1 + \varepsilon_2. \tag{6.85}$$

Defining $\lambda = \min\{\frac{2k_z}{M_{\max}}, \frac{2k_\varphi}{I_{\max}}\}$, we have

$$\dot{V}(t) \le -\lambda V + \varepsilon_1 + \varepsilon_2, \tag{6.86}$$

which shows that the Lyapunov function is bounded by

$$V(t) < V(0)e^{-\lambda t} - \frac{\varepsilon_1 + \varepsilon_2}{\lambda}(1 - e^{-\lambda t}). \tag{6.87}$$

Clearly, inequality (6.87) implies that the tracking errors p_z and p_φ are bounded, and the part A of the theorem is proven.

Proof (*Proof of B*) Choose a positive function as

$$V_u = V + \frac{1}{2}\sum_{i=1}^{2} r_{\alpha i}^{-1}\tilde{\alpha}_i^2 + \frac{1}{2}\sum_{i=1}^{2} r_{\beta i}^{-1}\tilde{\beta}_i^2 + \frac{1}{2}r_M^{-1}\tilde{M}^2 + \frac{1}{2}r_I^{-1}\tilde{I}^2,$$

and we have

$$\begin{aligned}
\dot{V}_u(t) &= \dot{V} + \sum_{i=1}^{2} r_{\alpha i}^{-1}\tilde{\alpha}_i\dot{\hat{\alpha}}_i + \sum_{i=1}^{2} r_{\beta i}^{-1}\tilde{\beta}_i\dot{\hat{\beta}}_i + r_M^{-1}\tilde{M}\dot{\hat{M}} + r_I^{-1}\tilde{I}\dot{\hat{I}} \\
&\le -k_z p_z^2 - k_\varphi p_\varphi^2 + \tilde{\alpha}_1(r_{\alpha 1}^{-1}\dot{\hat{\alpha}}_1 - (p_z + ap_\varphi)u_1^*) \\
&\quad + \tilde{\alpha}_2(r_{\alpha 2}^{-1}\dot{\hat{\alpha}}_2 - (p_z - bp_\varphi)u_2^*) + \tilde{\beta}_1(r_{\beta 1}^{-1}\dot{\hat{\beta}}_1 - p_z - ap_\varphi) \\
&\quad + \tilde{\beta}_2(r_{\beta 2}^{-1}\dot{\hat{\beta}}_2 - p_z + bp_\varphi) + \tilde{M}(r_M^{-1}\dot{\hat{M}} + p_z(\ddot{r}_z - k_z\dot{e}_z)) \\
&\quad + \tilde{I}(r_I^{-1}\dot{\hat{I}} + p_\varphi(\ddot{r}_\varphi - k_\varphi\dot{e}_\varphi)).
\end{aligned}$$

Noticing the two property of the projection mapping $Proj_\theta(r_\theta\tau_\theta)$, we have $\dot{V} \le -k_z p_z^2 - k_\varphi p_\varphi^2 \le 0$, which implies that the tracking errors p_z and p_φ will asymptotically converge to zero, by using barbalat's Lemma.

Table 6.3 The controller parameters of active suspensions

Parameter	$\gamma_{\alpha 1}$	$\gamma_{\alpha 2}$	γ_{β_1}	γ_{β_2}	γ_M	γ_I	k_z	k_φ	k_{pz}	$k_{p\varphi}$
Value	1	1	1	1	1000	66	100	10	1000	1000

6.3.3 Simulation Verification

In this subsection, we provide an example to illustrate the effectiveness of the proposed approach. The half-car model parameters are given as: $M = 1200$ kg, $M_{\max} = 1500$ kg, $M_{\min} = 1000$ kg, $I = 600$ kgm^2, $I_{\max} = 800$ kgm^2, $I_{\min} = 500$ kgm^2, $m_f = m_r = 100$ kg, $k_{f1} = k_{r1} = 15000$ N/m, $k_{nf1} = k_{nr1} = 1000$ N/m, $k_{f2} = 200000$ N/m, $k_{r2} = 150000$ N/m, $b_{f2} = b_e = 1500$ Ns/m, $b_{r2} = 2000$ Ns/m, $b_c = 1200$ Ns/m, $a = 1.2$ m, $b = 1.5$ m, $V_v = 30$ m/s, $\alpha_{i\max} = 1$, $\alpha_{i\min} = 0.5$, $\beta_{i\max} = 500$; $\beta_{i\min} = 0$, $i = 1, 2$.

Give the initial state values as: $x_1(0) = 6$ cm, $M(0) = 1500$, $I(0) = 500$, $\alpha_i(0) = 1$ and the rest ones are assumed as zeros. Besides, to investigate the effect of required settling time T_r on the system response, let the parameters of reference trajectory be $T_r = 0.5$ s; 1.0 s; 2.0 s, respectively. The controller parameters are given in Table 6.3.

In the following, we will successively illustrate the performances of the closed-loop suspension system in the case of no actuator fault and actuator fault. The proposed controller is tested by a road of the form of 3 Hz sine wave of 3 cm amplitude, and the sinusoidal road input is formulated as

$$z_o = h_1 \sin(6\pi t).$$

Here, it is assumed that the friction forces of suspension components F_l and F_φ are

$$F_l = F_\varphi = 50\sin(6\pi t)\text{N}.$$

Simulations Without Actuator Faults

Firstly, we verify the effectiveness of the proposed control laws when the active suspension systems subjected to no actuator faults. The time histories of vertical displacements for the passive suspension system, closed-loop active suspension system with different setting time T_r are compared in Fig. 6.11, in which the solid lines represent the responses of the active suspension systems with the proposed fault-tolerant controllers, and black, blue, and green lines are responses of closed-loop active suspension systems with $T_r = 0.5$, 1.0, 2.0 s, respectively. From this figure, it is seen that the magnitudes for the vertical motion are significantly decreased for active suspensions with proposed controller, compared with passive systems. Simultaneously, the settle time can be adjusted by giving the different values of T_r.

Figure 6.12 shows the corresponding curves of vertical accelerations. It can be seen that theoretically, we can settle T_r as arbitrarily small value to make the vertical or pitch displacements vanish fast. However, smaller settle time T_r results in larger

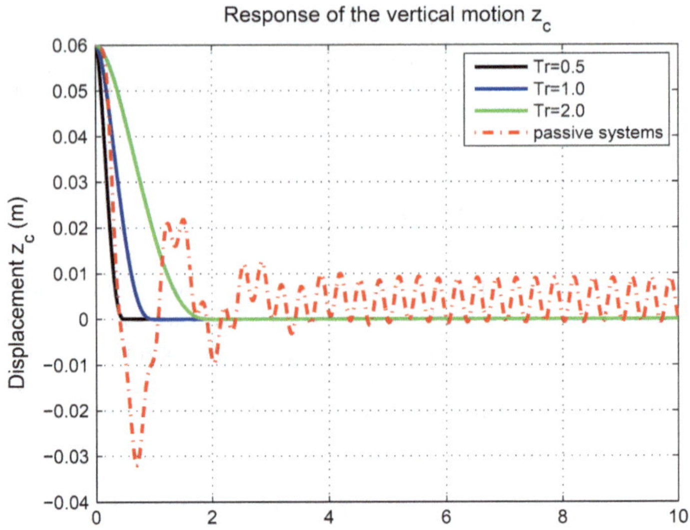

Fig. 6.11 Vertical displacements of car body (no actuator faults)

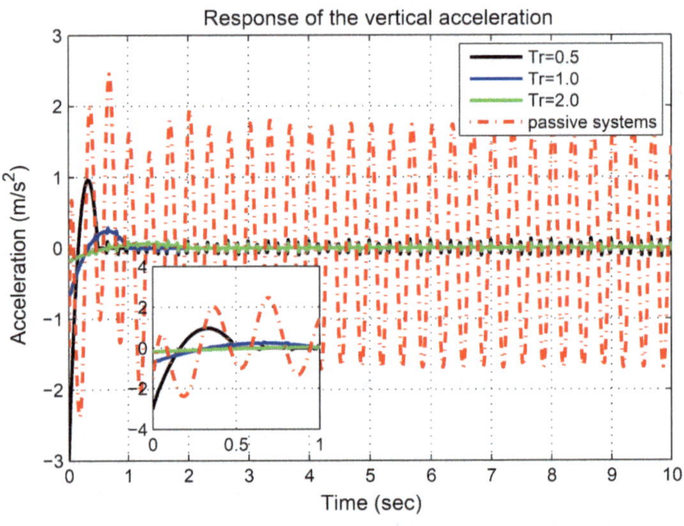

Fig. 6.12 Vertical acceleration of car body (no actuator faults)

vertical and pitch accelerations and larger tracking errors e_1, which can be confirmed from Fig. 6.13.

The responses of pitch angle displacements, accelerations and tracking errors are given in Figs. 6.14, 6.15 and 6.16. For pitch motion, the reference trajectory is chosen as "zero", and the aim behind this is to stabilize the pitch motion as soon as possible. From Figs. 6.14 and 6.15, we can see that our proposed approaches can

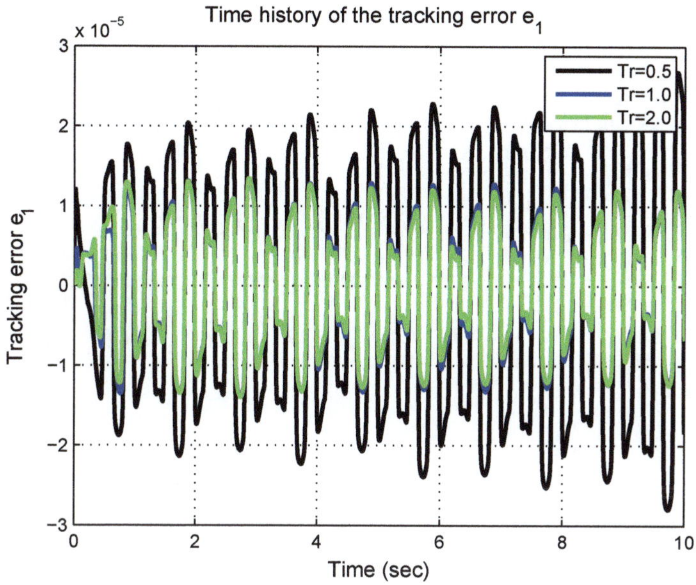

Fig. 6.13 Tracking errors e_1 (no actuator faults)

yield smaller peak values and shorter settling time, compared with passive systems, and the maximum tracking error is just about 1×10^{-4} which is very helpful to improve the ride comfort.

From the point of view of car safety, the firm uninterrupted contact of wheels to road should be ensured. The dynamic tire load behavior is one of the important properties of the suspensions, due to its relation with the road holding. Here, the performance of road holding is considered here based on an idea constraint for road handling, that is, the dynamic tire load should not exceed the static ones for both of the front and rear wheels, i.e.

$$\left|D_f\right| = \left|F_{tf} + F_{bf}\right| < F_f, \quad |D_r| = |F_{tr} + F_{br}| < F_r, \tag{6.88}$$

where the static tire loads F_f and F_r are computed by

$$\begin{aligned} F_f + F_r &= (M_{\min} + m_f + m_r)g, \\ F_f(a+b) &= M_{\min}gb + m_f g(a+b). \end{aligned} \tag{6.89}$$

Then the ride safety index can be defined as

$$\frac{|F_{ti} + F_{bi}|}{F_i}, i = f, r,$$

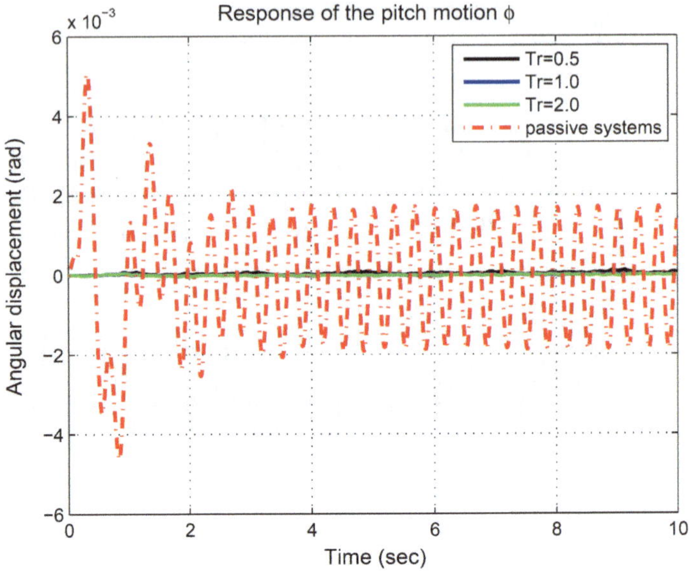

Fig. 6.14 Pitch angle displacements of car body (no actuator faults)

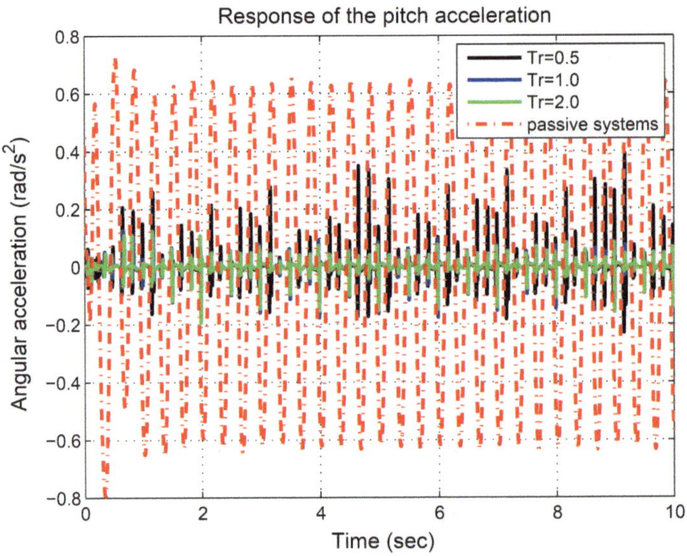

Fig. 6.15 Pitch angle acceleration of car body (no actuator faults)

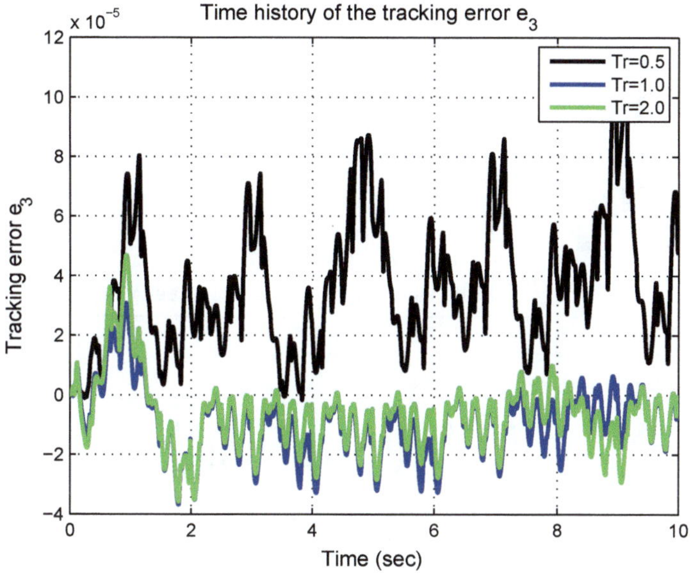

Fig. 6.16 Tracking errors e_3 (no actuator faults)

which means that if this index is less than 1, then the ride safety performance can be guaranteed. The static tire loads for both front and rear wheels can be calculated as $F_f = 7513.3\,\text{N}$, $F_r = 6206.7\,\text{N}$. Figure 6.17 shows that the ride safety indexes for both the front and rear wheels are less than 1, which ensures the ride safety.

In the active suspension control, the limitations of the suspension space should be taken into account, which means the suspension working space must be preserved. It can been observed from Fig. 6.18 that the controlled suspension spaces all fall into the acceptable ranges, whose maximums are $z_{\max} = 0.1$ m. Figure 6.19 are plotted to show the actuator forces.

Simulations with Actuator Faults

From the simulation results previously, it can be seen that the proposed fault-tolerant controllers can achieve good performances, in the case that there is no actuator faults. In this part, the fault-tolerant ability of the proposed controllers will be verified. Firstly, choose $T_r = 2$ s, and assume that the front actuator u_1 gets totally broken ($\bar{u}_1 = 0$ N) at $t = 6$ s (fault 1), and the actuators then can be represented as

$$u_1 = \begin{cases} u_{1s}^*, & t < 6\text{s} \\ 0, & t \geq 6\text{s} \end{cases}, \quad u_2 = u_{2s}^*.$$

For comparison purposes, three systems are defined as follows:

SA: The proposed fault-tolerant controllers to control the active suspension systems with actuator faults (System A);

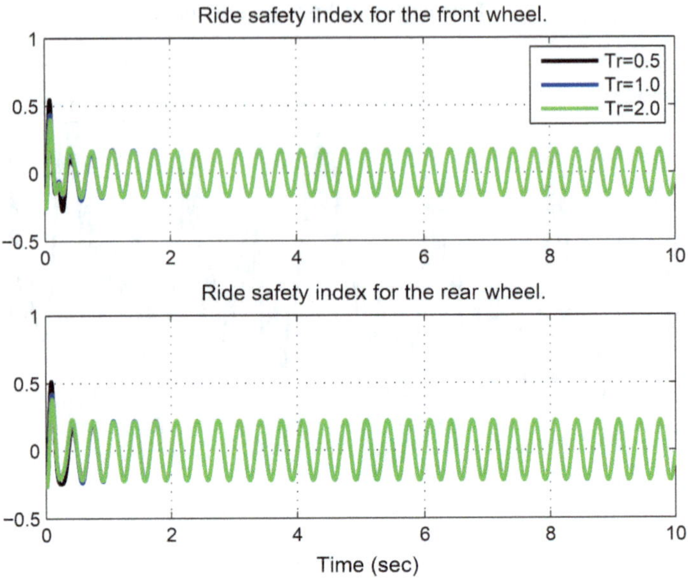

Fig. 6.17 Ride safety indexes (no actuator faults)

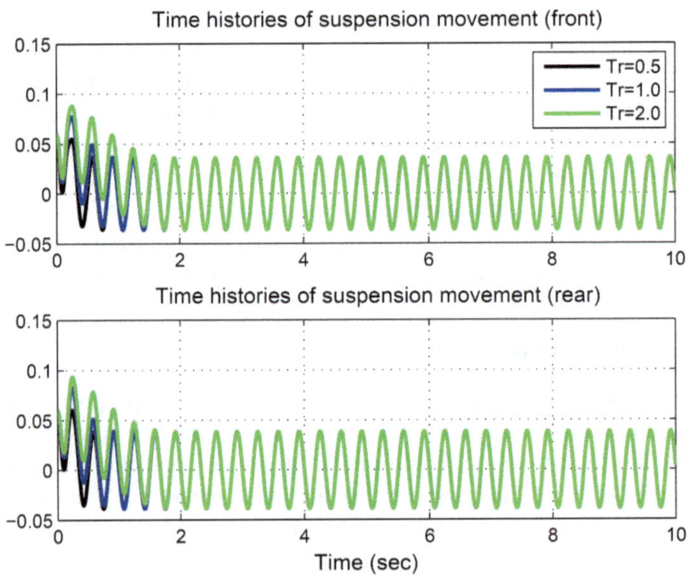

Fig. 6.18 Suspension movements (no actuator faults)

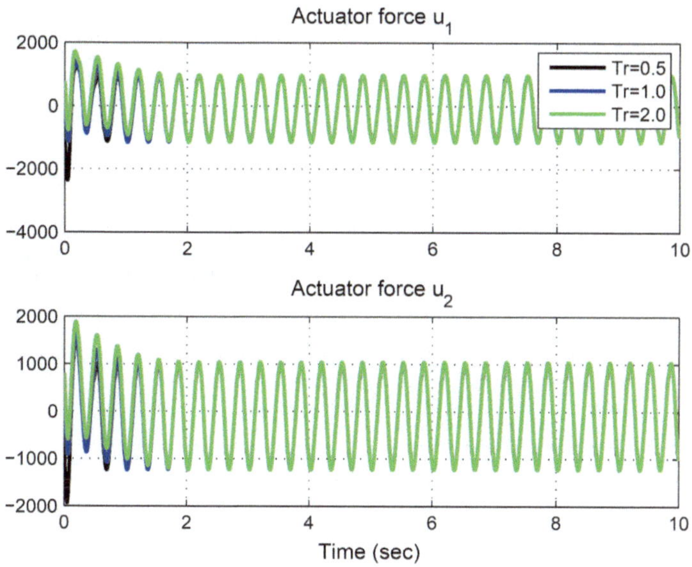

Fig. 6.19 Actuator forces (no actuator faults)

SB: The proposed fault-tolerant controllers to control the active suspension systems without actuator faults (System B);

SC: The traditional ARC controllers to control the active suspension systems with actuator faults (System C).

Figures 6.20 and 6.21 plot the vertical and pitch angle displacements for the four closed-loop systems, from which we can see that "System A" yields a better performance than "System C", especially for the pitch motion, which means that the proposed fault-tolerant controllers work well for the systems with actuator faults.

The ride safety indexes and suspension movements are given in Figs. 6.22 and 6.23, and all the curves in these two figures are below 1, which shows the constraints are guaranteed. Actuator forces are plotted in Fig. 6.24.

Next, assuming the rear actuator u_2 loses 40% of efficiency at 6s (fault 2), the actuators then can be represented as

$$u_1 = u_{1s}^*,\ u_2 = \begin{cases} u_{2s}^*, & t < 6\text{s} \\ 0.6u_{2s}^*, & t \geq 6\text{s} \end{cases}.$$

From Figs. 6.25, 6.26, 6.27, 6.28 and 6.29, it can be seen that the proposed fault-tolerant controllers can achieve almost the same performance with the healthy actuator even though actuator failures happen.

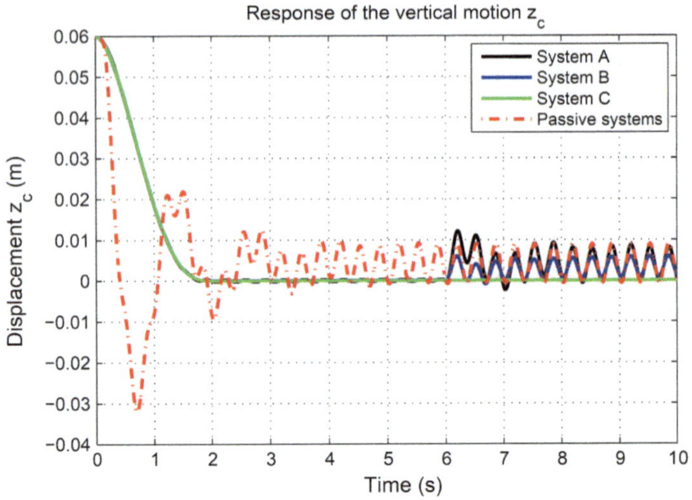

Fig. 6.20 Vertical displacements of car body (fault 1)

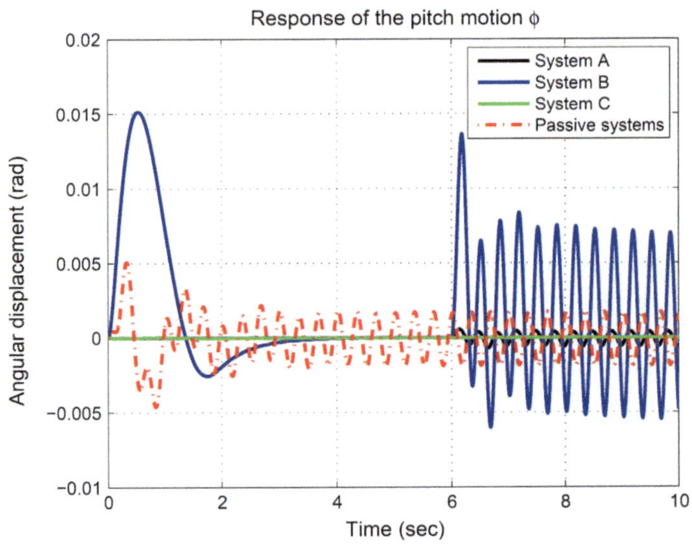

Fig. 6.21 Pitch angle displacements of car body (fault 1)

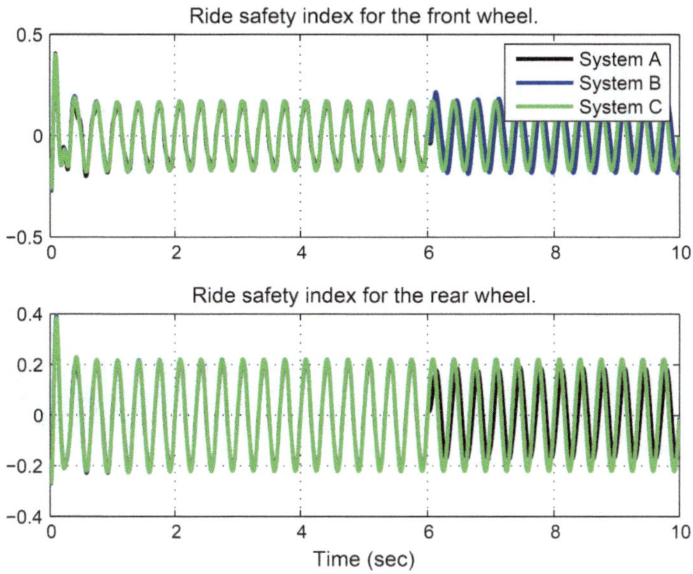

Fig. 6.22 Ride safety indexes (fault 1)

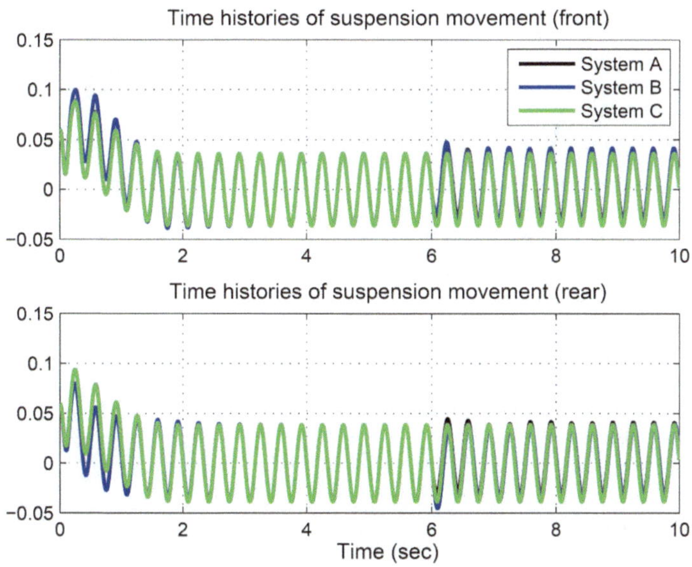

Fig. 6.23 Suspension movements (fault 1)

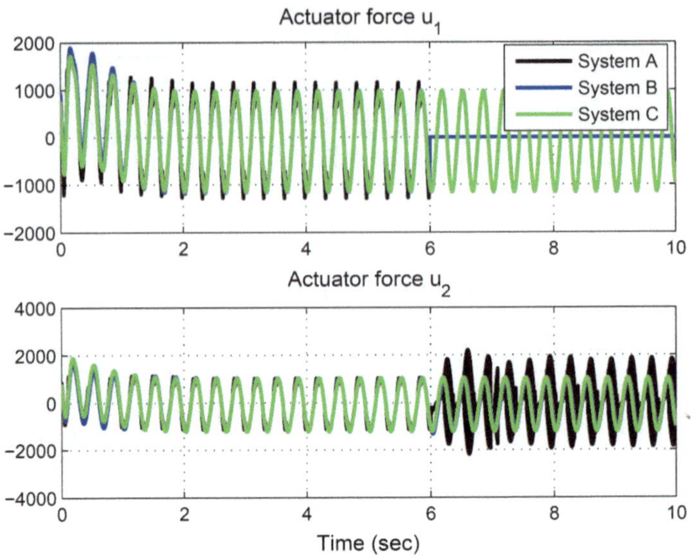

Fig. 6.24 Actuator forces (fault 1)

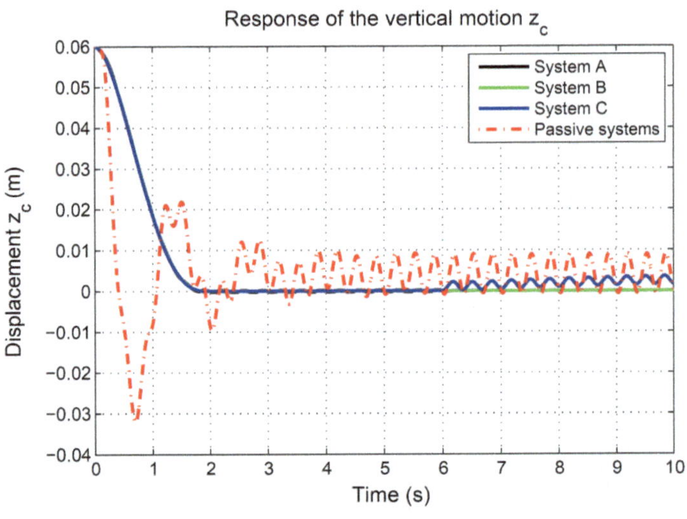

Fig. 6.25 Vertical displacements of car body (fault 2)

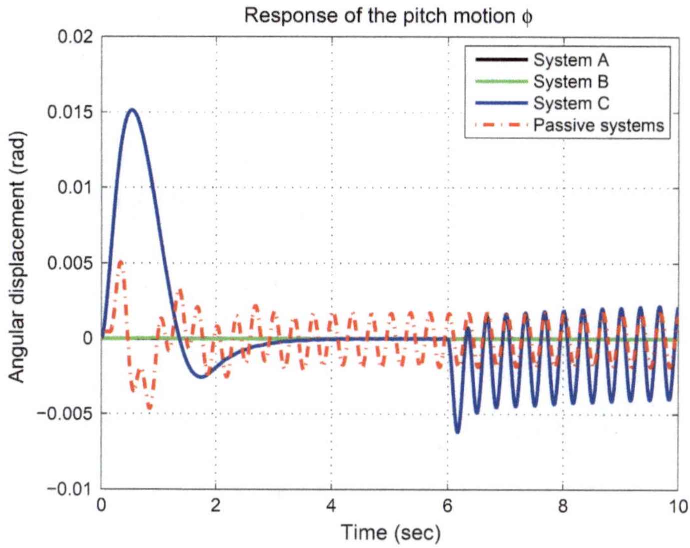

Fig. 6.26 Pitch angle displacements of car body (fault 2)

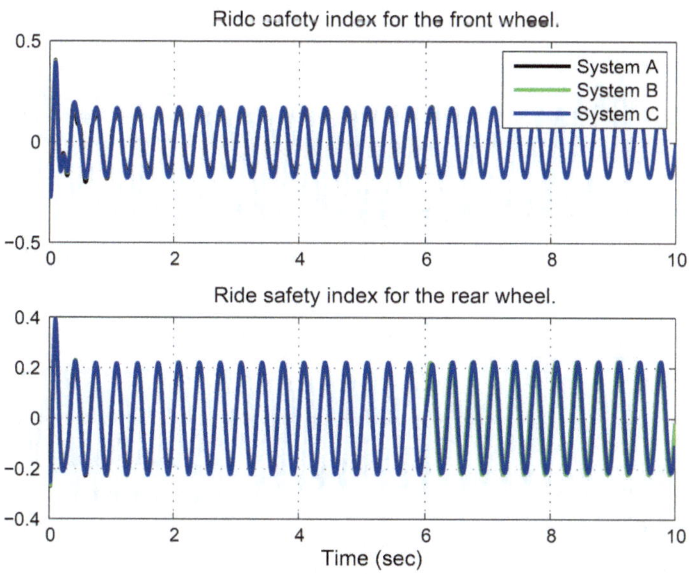

Fig. 6.27 Ride safety indexes (fault 2)

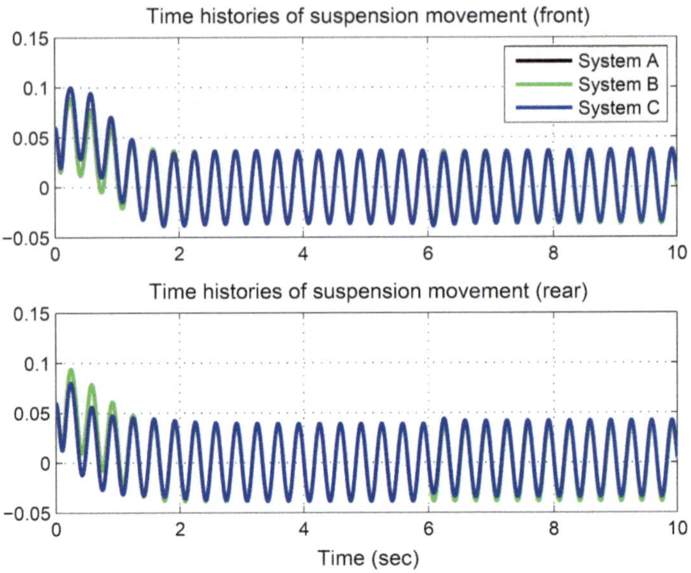

Fig. 6.28 Suspension movements (fault 2)

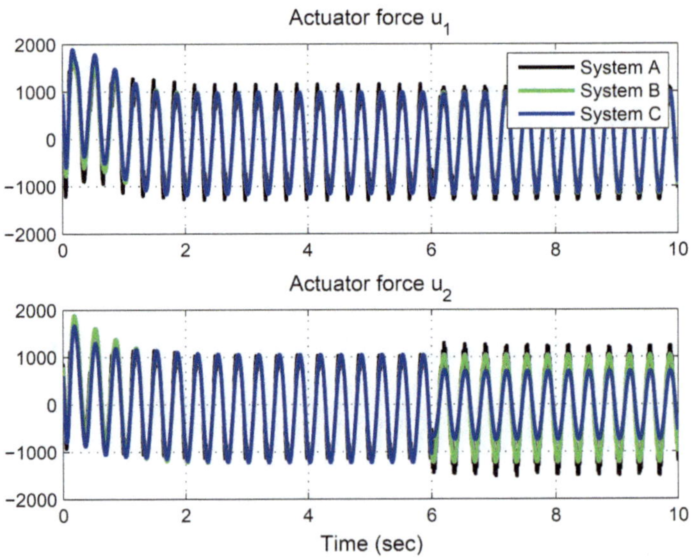

Fig. 6.29 Actuator forces (fault 2)

6.3.4 Conclusion

In this section, a fault tolerant control approach was proposed to deal with the problem of fault accommodation for unknown actuator failures of active suspension systems, where an adaptive robust controller was designed to adapt and compensate the parameter uncertainties, external disturbances and uncertain nonlinearities generated by the system itself and actuator failures. Comparative simulation studies were then given to illustrate the effectiveness of the proposed controllers.

As future works, the following interesting topics should be considered: firstly, implementing practical experiment to verify the fault tolerant control scheme is an important part, which can introduce the proposed control law into the practical engineering. In addition, considering the integrated fault tolerant control of the several subsystems (such as the active suspension systems, active front steering, anti-locked braking system, and so on) to improve the vehicle dynamics performances is one of important topics that need to be dealt with.

References

1. S. Tong, T. Wang, W. Zhang, Fault tolerant control for uncertain fuzzy systems with actuator failures. Int. J. Innov. Comput. Inf. Control **4**(10), 2461–2474 (2008)
2. T. Iwasaki, S. Hara, Generalized KYP lemma: unified frequency domain inequalities with design applications. IEEE Trans. Autom. Control **50**(1), 41–59 (2005)
3. H. Gao, W. Sun, P. Shi, Robust sampled-data H_∞ control for vehicle active suspension systems. IEEE Trans. Control Syst. Technol. **18**(1), 238–245 (2010)
4. H. Du, N. Zhang, H_∞ control of active vehicle suspensions with actuator time delay. J. Sound Vib. **301**, 236–252 (2007)
5. H. Chen, K. Guo, Constrained H_∞ control of active suspensions: an LMI approach. IEEE Trans. Control Syst. Technol. **13**(10), 412–421 (2005)
6. T. Gordon, C. Marsh, M. Milsted, A comparision of adaptive LQG and nonlinear controllers for vehicle suspension systems
7. S. Gayaka, B. Yao, Output feedback based adaptive robust fault-tolerant control for A class of uncertain nonlinear systems. J. Syst. Eng. Electron. **22**(1), 38–51 (2011)
8. J. Boskovic, R. Mehra, A decentralized fault-tolerant control system for accommodation of failures in higher-order flight control actuators. IEEE Trans. Control Syst. Technol. **18**(5), 1103–1115 (2010)
9. B. Yao, F. Bu, J. Reedy, G. Chiu, Adaptive robust motion control of single-rod hydraulic actuators: theory and experiments. IEEE Trans. Mechatron. **5**(1), 79–91 (2000)
10. B. Yao, G. Chiu, J. Reedy, Nonlinear adaptive robust control of one-DOF electro-hydraulic servo systems, in *ASME International Mechanical Engineering Congress and Exposition (IMECE'SY)* (1997)
11. W. Sun, H. Gao, O. Kaynak, Adaptive backstepping control for active suspension systems with hard constraints. IEEE/ASME Trans. Mechatron. **18**(3), 072–1079 (2013)

Chapter 7
Active Suspensions Control with Actuator Dynamics

In the previous chapters, the dynamic process of the actuator is neglected by us. In this chapter, we mainly look at actuator dynamics when designing control scheme, and electro-hydraulic systems are typically chosen as the actuators. However, the main drawback of electro-hydraulic systems is their highly non-linear behavior. In addition, there exists model uncertainties when establishing the mathematical model. All these factors make it difficult to design the controller and influence the effectiveness of active force tracking. To solve these problems, in Sect. 7.1, with regard to nonlinear half-car model, a filter-based adaptive vibration control scheme is suggested. Furthermore, in Sect. 7.2, an adaptive robust vibration control scheme of full-car active suspensions is proposed.

Nomenclature: The following nomenclature is used throughout the paper: $\hat{\bullet}$ is used to denote the estimate of $\bullet$, $\tilde{\bullet}$ is used to denote the parameter estimation error of $\bullet$, and $\bullet_{\max}$, $\bullet_{\min}$ are the maximum and minimum value of $\bullet(t)$ for all t, respectively. For a matrix $\bullet$, $\bullet^T$ denotes its transpose; the notation $\bullet > 0$ (≥ 0) means that $\bullet$ is real symmetric and positive definite (semi-definite). In this study, $\|\bullet\|_\infty$ denotes the ∞-norm, which obeys $\|x\|_\infty = \max(x_j),\ j = 1, \ldots, n$.. For a matrix P, P^T and P^{-1} denote its transpose and inverse, respectively; the notation $P > 0$ (≥ 0) means that P is real symmetric and positive definite (semi-definite). For a vector or matrix, $\{\cdot\}_j$ $(j = 1, 2, \ldots)$ represents the jth line of the vector or matrix, and $\|\cdot\|_\infty$, $\|\cdot\|$ and $\|\cdot\|_2$ denote the ∞-norm, Euclidian norm and two-norm of the vector. In symmetric block matrices or complex matrix expressions, we use an asterisk $(*)$ to represent a term that is induced by symmetry and diag$\{\ldots\}$ stands for a block-diagonal matrix. Matrices, if their dimensions are not explicitly stated, are assumed to be compatible for algebraic operations.

W. Sun et al., *Advanced Control for Vehicle Active Suspension Systems*, Studies in Systems, Decision and Control 204,
https://doi.org/10.1007/978-3-030-15785-2_7

7.1 Filter-Based Adaptive Vibration Control for Active Suspensions with Electro-Hydraulic Actuators

In this section, an adaptive vibration control strategy is proposed for the nonlinear uncertain suspension systems to stabilize both the vertical and pitch motions of the car, and thus to contribute to the ride comfort. Simultaneously, the ride holding performances are preserved within their allowable limits in the controller design. Moreover, differing from the existing results, in most of which the effect of actuator dynamic is neglected, this section considers the electro-hydraulic systems as actuators to supply the active forces into suspension systems. Furthermore, to overcome the "exploration of terms" problem existing in standard backstepping, a filter-based adaptive control strategy is subsequently proposed. Finally, a design example is shown to illustrate the effectiveness of the proposed active controllers, where different road conditions are considered in order to reveal the closed-loop system performance in details.

7.1.1 *Problem Formulation*

The effect of actuator dynamics is often neglected and the actuators are modeled as ideal force generators. Such assumption results in inaccuracies of controller design in actual engineering.

In active suspensions design, electro-hydraulic systems are typically chosen as the actuators to generate the forces to isolate the vibrations transmitted to the passengers. This is because they are more powerful and less bulky compared with other actuators. However, the main drawback of electro-hydraulic systems is their highly non-linear behavior, making it more difficult to design the ideal control law. The classical and commonly used approach in the control of electro-hydraulic servo systems is based on local linearization of the nonlinear dynamics of the system [1]. Such approach requires conservative controllers that sacrifices performance in favour of robustness. Reference [2] uses a Linear Parameter Varying (LPV) control method to control an injection-moulding machine. Nevertheless, due to their highly nonlinear dynamics, using electrohydraulic actuators to track the desired forces is fundamentally limited when interacting with a dynamic environment [3]. Techniques such as adaptive fuzzy sliding control [4] was then proposed to solve this problem. However, these approaches need a complicated learning mechanism, which is designed by a trial and error process, and presents certain difficulties in application.

Another problem to be considered is parametric uncertainties in system modeling. There are several sources to result in the parametric uncertainties. (1) Because of the change of the number of passengers or the payload, vehicle load is easily varied, which will accordingly change the vehicle mass as a varying parameter; (2) the change of vehicle mass certainly leads to the variation of the moment of inertia; (3) in electro-hydraulic actuator systems, the effective bulk modulus β_e is usually an unknown parameter, whose range is 200–800 Mpa. These uncertain parameters will bring the enormous difficulties for controller design.

Half-Car active suspension systems with electro-hydraulic actuators are considered as the control plant, and based on this model, we propose a constrained adaptive backstepping strategy to stabilize both the vertical and pitch motions, despite there exist uncertain parameters and highly nonlinearities. Simultaneously, ride safety constraints required by advanced vehicle suspension can be guaranteed in the controller design by adjusting the tuning parameters and initial values.

Nonlinear Active Suspension Model: Half-Car Model

In this section, the nonlinear half-car model shown in Fig. 1.3 in Sect. 1.2.2 is considered. This model has been used extensively in the literature and captures many important characteristics of vertical and pitch motions. Here, the effect of lateral motion is neglected. The ideal dynamic equations of the sprung and unsprung masses are given by (4.43) in Sect. 4.2. k_{s1}, k_{s2} and b_{e1}, b_{e2} are the stiffness and damping coefficients of the suspension system, respectively, and k_{t1}, k_{t2} and b_{t1}, b_{t2} stand for compressibleness and damping coefficients of the pneumatic tyre, respectively. And F_1, F_2 in place of u_1, u_2 are the forces from electro-hydraulic actuators.

Tire Subsystems

Tire subsystems in vehicle can be simplified as a spring-damper-mass system, whose dynamic equations can be written as:

$$\mathbf{M_T}: \begin{cases} m_f\ddot{z}_1 - k_{s1}\Delta y_1 - b_{e1}\Delta\dot{y}_1 + k_{t1}(z_1 - z_{o1}) \\ \qquad\qquad\qquad + b_{t1}(\dot{z}_1 - \dot{z}_{o1}) = -F_1, \\ m_r\ddot{z}_2 - k_{s2}\Delta y_2 - b_{e2}\Delta\dot{y}_2 + k_{t2}(z_2 - z_{o2}) \\ \qquad\qquad\qquad + b_{t2}(\dot{z}_2 - \dot{z}_{o2}) = -F_2, \end{cases} \tag{7.1}$$

where Δy_i and $\Delta\dot{y}_i$, $i = 1, 2$, represent the suspension displacement and velocity with the form of

$$\begin{aligned} \Delta y_1 &= z_c + a\sin\varphi - z_1, \quad \Delta y_2 = z_c - b\sin\varphi - z_2, \\ \Delta\dot{y}_1 &= \dot{z}_c + a\dot{\varphi}\cos\varphi - \dot{z}_1, \quad \Delta\dot{y}_2 = \dot{z}_c - b\dot{\varphi}\cos\varphi - \dot{z}_2. \end{aligned}$$

Active Suspension Subsystems

Active suspension subsystems basically consist of springs, shock absorbers and actuators to transmit and filter all forces between body and road. The dynamic equations of active suspension subsystems are given as:

$$\mathbf{M_S}: \begin{cases} M\ddot{z}_c + \phi_1(t) = F_1 + F_2, \\ I\ddot{\varphi} + \phi_2(t) = aF_1 - bF_2, \end{cases} \tag{7.2}$$

where

$$\begin{aligned}\phi_1(t) &= k_{s1}\Delta y_1 + k_{s2}\Delta y_2 + b_{e1}\Delta \dot{y}_1 + b_{e2}\Delta \dot{y}_2,\\ \phi_2(t) &= a(k_{s1}\Delta y_1 + b_{e1}\Delta \dot{y}_1) - b(k_{s2}\Delta y_2 + b_{e2}\Delta \dot{y}_2).\end{aligned}$$

Due to the change of the number of passengers or the payload, vehicle mass M is easily varied, and thus, the moment of inertia I is varied correspondingly.

Actuator Subsystems

In the study, electro-hydraulic systems are chosen as actuators for active control, and their dynamics is taken into consideration as shown in Fig. 7.1. Here, the valve dynamics are neglected and the servo-valve opening x_{vi} is directly related to the control input by a known static mapping, that is $x_{vi} = k_{vi}u_i$, $i = 1, 2$. Neglecting the effect of leakage flows in the cylinder, the cylinder dynamics can be written as

$$\dot{P}_{Li} = -\beta P_{Li} - \alpha A \Delta \dot{y}_i + \mu x_{vi} g_i(\cdot),\ i = 1, 2,$$

where $P_{Li} = P_{1i} - P_{2i}$ is the load pressure of the cylinder, A is the ram area of the cylinder, $\alpha = \frac{4\beta_e}{V_t}$, $\beta = \alpha C_{tp}$, $\mu = \alpha C_d \omega \sqrt{\frac{1}{\rho}}$, $g_i(\cdot) = \sqrt{P_s - \text{sgn}(x_{vi})P_{Li}}$, β_e is the effective bulk modulus, V_t is the total actuator volume, C_{tp} is the total leakage coefficient of the piston, ρ is the hydraulic fluid density, ω is the spool valve area gradient, C_d is the discharge coefficient and P_s is supply pressure. Noting that the forces generated by actuators can be described as $F_i = AP_{Li}$, the force dynamics of electro-hydraulic systems are given as

$$\mathbf{M_A} : \dot{F}_i = -\beta F_i - \alpha A^2 \Delta \dot{y}_i + A\mu x_{vi} g_i(\cdot),\ i = 1, 2. \tag{7.3}$$

To facilitate study follow-up, we first give the following assumption:

Assumption 6 It is assumed that

$$\begin{aligned} M &\in \Omega_M = \{M : M_{\min} \le M \le M_{\max}\},\\ I &\in \Omega_I = \{I : I_{\min} \le I \le I_{\max}\},\end{aligned}$$

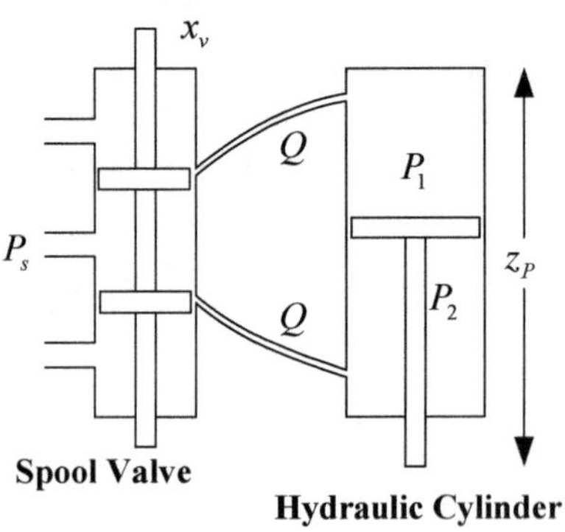

Fig. 7.1 The schematic diagram of electro-hydraulic actuator

where $M_{\min}$, $I_{\min}$ and $M_{\max}$, $I_{\max}$ stand for the lower and upper bounds of vehicle mass and the moment of inertia, respectively.

Assumption 7 For electro-hydraulic systems, it is assumed that β_{ei} is an uncertain parameter which satisfies

$$\beta_{ei} \in \Omega_\beta = \{\beta_{ei} : \beta_{ei\,\min} \le \beta_{ei} \le \beta_{ei\,\max}\},$$

where $\beta_{ei\,\min}$ and $\beta_{ei\,\max}$ are the lower and upper bounds of β_{ei}, $i = 1, 2$.

Remark 7.1 Assumptions 1 and 2 are rather weak as it is easy to obtain the upper and lower bounds of unknown parameters M, I, β_{e1} and β_{e2} based on the real physical systems.

Problem Statement

Based on the electro-hydraulic active suspension model, we aim to keep the vertical and pitch motions (z_c and φ) to converge to zero or be bounded, and simultaneously, ride safety constraints are guaranteed within their allowable range. The control problems can be described as follows.

Problem 7.2 Synthesize a constrained adaptive backstepping control strategy u such that

$$\lim_{t\to\infty} z_c \to 0 \text{ or bounded}, \quad \lim_{t\to\infty} \varphi \to 0 \text{ or bounded}.$$

Additionally, to guarantee the road holding, the dynamic tire load should not exceed the static ones for both of the front and rear wheels, and thus, the following performance constraints are guaranteed:

$$\left| \frac{k_{ti}(z_i - z_{oi}) + b_{ti}(\dot{z}_i - \dot{z}_{oi})}{S_i} \right| < 1, \ i = 1, 2, \tag{7.4}$$

where S_1 and S_2 are the static tire loads which can be computed by

$$\begin{aligned} S_1 + S_2 &= (M + m_f + m_r)g, \\ S_1(a + b) &= Mgb + m_f g(a + b). \end{aligned} \tag{7.5}$$

7.1.2 Adaptive Backstepping Controller Synthesis

Controller Design

Step 1: In this step, our target is to find a desired function F_z^d such as if $F_1 + F_2 = F_z^d$, then the tracking error z_c can converge to zero.

Defining $p_z = \dot{z}_c + k_z z_c$, where $k_z > 0$ is a constant, it can be known that z_c will converge to zero or be bounded as long as p_z converges to zero or be bounded, because the transfer function $\frac{z_x(s)}{p_z(s)} = \frac{1}{s+k_z}$ is stable [5, 6]. Then, the control target in this step switches to keep p_z converge to zero or be bounded.

Firstly, we have

$$M\dot{p}_z = M\ddot{z}_c + k_z M\dot{z}_c = F_1 + F_2 - \phi_1(t) + k_z M\dot{z}_c,$$

then we can design a desired function F_z^d and an adaptive law of body mass M as

$$F_z^d = \phi_1(t) - k_z \hat{M}\dot{z}_c - k_{pz} p_z, \quad \dot{\hat{M}} = r_M k_z \dot{z}_c p_z, \tag{7.6}$$

to guarantee $p_z \to 0$ or be bounded when $F_1 + F_2 = F_z^d$, where $k_{pz} > 0$ and $k_z > 0$ are constants, and $r_M > 0$ is the adaptive gain.

Step 2: Design a desired function F_φ^d such as if $aF_1 - bF_2 = F_\varphi^d$, then the pitch angel φ can converge to zero or be bounded.

Following the same line with "step 1", we define $p_\varphi = \dot{\varphi} + k_\varphi \varphi$, where $k_\varphi > 0$ is a constant. Afterwards, we have

$$I\dot{p}_\varphi = I\ddot{\varphi} + k_\varphi I\dot{\varphi} = aF_1 - bF_2 - \phi_2(t) + k_\varphi I\dot{\varphi},$$

Finally, we can get the conclusion of

$$F_\varphi^d = \phi_2(t) - k_\varphi \hat{I}\dot{\varphi} - k_{p\varphi} p_\varphi, \quad \dot{\hat{I}} = r_I k_\varphi \dot{\varphi} p_\varphi, \tag{7.7}$$

where $k_{p\varphi} > 0$ is a constant, and $r_I > 0$ is the adaptive gain. Similarly, if $aF_1 - bF_2 = F_\varphi^d$, then p_φ will converge to zero or be bounded, and thus, the pitch angle φ will converge to zero or be bounded.

After obtaining F_z^d and F_φ^d, we can get the desired output forces of electro-hydraulic actuators as

$$F_1^d = \frac{bF_z^d + F_\varphi^d}{a+b}, \quad F_2^d = \frac{aF_z^d - F_\varphi^d}{a+b}.$$

If F_i is equal to F_i^d, $i = 1, 2$, then the targets of "step 1 and step 2" can be achieved. However, F_i, $i = 1, 2$, are not the real inputs, but only the states, so we can not assign the desired functions F_i^d to F_i ($i = 1, 2$) directly. This fact leads to the design of next step.

Step 3: Design the real inputs u_i to keep F_i track F_i^d ($i = 1, 2$) as close as possible.

To work out this problem, it is necessary to calculate the derivative of F_i^d in traditional backstepping type approach, but as we know, sometimes, it is difficult to do this work. Here, a filter-based approach is employed to estimate $\dot{F}_i^d$ and thus to avoid accelerating it [7, 8], that is

$$\tau_i \dot{\bar{F}}_i^d + \bar{F}_i^d = F_i^d, \quad \bar{F}_i^d(0) = F_i^d(0), \; i = 1, 2,$$

where $\tau_i > 0$ is time constant of the filter. Given the definitions of $y_{F_i} = \bar{F}_i^d - F_i^d$ and $e_{F_i} = F_i - \bar{F}_i^d$, we have

$$\dot{e}_{F_i} = \dot{F}_i - \dot{\bar{F}}_i^d = \beta_{ei}\left(-\frac{4C_{tp}}{V_t}F_i - \frac{4A^2}{V_t}\Delta\dot{y}_i + Q_i\right) + \frac{y_{F_i}}{\tau_i}, \tag{7.8}$$

where $Q_i = \frac{4AC_d\omega}{V_t\sqrt{\rho}}x_{vi}g_i(\cdot)$, $i = 1, 2$. Based on the error dynamics (7.8), we can design

$$\begin{aligned}
Q_1 &= \frac{1}{\hat{\beta}_{e1}}\left(-p_z - ap_\varphi - \frac{y_{F_1}}{\tau_1} - k_{e1}e_{F_1}\right) + \frac{4C_{tp}}{V_t}F_1 + \frac{4A^2}{V_t}\Delta\dot{y}_1,\\
Q_2 &= \frac{1}{\hat{\beta}_{e2}}\left(-p_z + bp_\varphi - \frac{y_{F_2}}{\tau_2} - k_{e2}e_{F_2}\right) + \frac{4C_{tp}}{V_t}F_2 + \frac{4A^2}{V_t}\Delta\dot{y}_2,
\end{aligned}$$

and the adaptation of unknown parameters

$$\dot{\hat{\beta}}_{ei} = r_i\Theta_i(\cdot)e_{F_i},\ i = 1, 2, \tag{7.9}$$

where r_i and k_{ei} are positive constants, and

$$\Theta_i(\cdot) = -\frac{4C_{tp}}{V_t}F_i - \frac{4A^2}{V_t}\Delta\dot{y}_i + Q_i.$$

Finally, the original control signal $x_{vi} = k_{vi}u_i$ ($i = 1, 2$) can then be found from the definition of Q_i, i.e.

$$u_i = \frac{V_tQ_i}{4k_{vi}AC_d\omega\sqrt{\frac{P_s - \mathrm{sgn}(x_{vi})P_{Li}}{\rho}}},\ i = 1, 2. \tag{7.10}$$

Stability Analysis of the Closed-Loop Systems

Based on the previous design, we can first give the following closed-loop error systems

$$\begin{aligned}
M\dot{p}_z &= \sum_{i=1}^{2}(e_{F_i} + y_{F_i}) - k_{pz}p_z - k_z\tilde{M}\ddot{z}_c,\\
I\dot{p}_\varphi &= a(e_{F_1} + y_{F_1}) - b(e_{F_2} + y_{F_2}) - k_{p\varphi}p_\varphi - k_\varphi\tilde{I}\dot{\varphi},\\
\dot{e}_{F_1} &= -p_z - ap_\varphi - k_{e1}e_{F_1} - \tilde{\beta}_{e1}\Theta_1(\cdot),\\
\dot{e}_{F_2} &= -p_z + bp_\varphi - k_{e2}e_{F_2} - \tilde{\beta}_{e2}\Theta_2(\cdot).
\end{aligned}$$

Before showing the main result of stability, the sets and values which will be used in the stability proof are defined below.

For any $\sigma > 0$, define

$$\Pi = \left\{(p_z, p_\varphi, \tilde{M}, \tilde{I}, e_{F_i}, \tilde{\beta}_{ei}, y_{F_i}, i = 1, 2) : V(t) \le \sigma\right\}$$

where

$$\begin{aligned} V(t) =& \frac{1}{2} M p_z^2 + \frac{1}{2} I p_\varphi^2 + \frac{1}{2} r_M^{-1} \tilde{M}^2 + \frac{1}{2} r_I^{-1} \tilde{I}^2 \\ &+ \frac{1}{2} \sum_{i=1}^{2} (e_{F_i}^2 + r_i^{-1} \tilde{\beta}_{ei}^2 + y_{F_i}^2). \end{aligned} \tag{7.11}$$

Noting that

$$y_{F_i} = \bar{F}_i^d - F_i^d,$$

the derivative of y_{F_i} is

$$\dot{y}_{F_i} = -\frac{y_{F_i}}{\tau_i} - \dot{F}_i^d, \ i = 1, 2. \tag{7.12}$$

All terms in (7.12) can be dominated by some continuous functions, therefore, we have

$$\left| \dot{y}_{F_i} + \frac{y_{F_i}}{\tau_i} \right| \le B_i(t), \ i = 1, 2, \tag{7.13}$$

where $B_i(t)$ stands for continuous functions. Then, the following inequality can be obtained:

$$y_{F_i} \dot{y}_{F_i} + \frac{y_{F_i}^2}{\tau_i} \le \left| y_{F_i} \right| \left| \dot{y}_{F_i} + \frac{y_{F_i}}{\tau_i} \right| \le B_i(t) \left| y_{F_i} \right|. \tag{7.14}$$

Thus, we have

$$y_{F_i} \dot{y}_{F_i} \le B_i(t) \left| y_{F_i} \right| - \frac{y_{F_i}^2}{\tau_i} \le -\frac{y_{F_i}^2}{\tau_i} + y_{F_i}^2 + \frac{1}{4} B_i^2(t). \tag{7.15}$$

Obviously, Π is a compact subset in **R,** hence there must be a point corresponding to the supreme value of $B_i(t)$ in Π. We denote this supreme value as Δ_i, that is

$$B_i(t) \le \Delta_i, \ i = 1, 2.$$

The following theorem is given to conclude the main result:

Theorem 7.3 *Considering the nonlinear active suspension system, if the control law is given as (7.10), and adaptive laws are given as (7.6), (7.7) and (7.9), then for any initial states in Π, there exist positive parameters k_{pz}, $k_{p\varphi}$, k_{ei},and τ_i, $i = 1, 2$, satisfying $\exists \lambda > 0$,*

$$\sum\nolimits_p \begin{cases} k_{pz} - \frac{1}{2} \geq M_{\max}\lambda, \\ k_{p\varphi} - \frac{1}{2}a^2 - \frac{1}{2}b^2 \geq I_{\max}\lambda, \\ k_{ei} \geq \lambda, \\ \frac{1}{\tau_i} - 2 \geq \lambda, \ i = 1, 2, \end{cases}$$

such that:

(1) all the error signals $(p_z, \ p_\varphi, \ e_{F_1}, \ e_{F_2})$ *in the closed-loop error systems are uniformly ultimately bounded and the steady-state tracking errors can be arbitrarily small. Especially, due to the stable transfer functions form* $p_z(s)$ *and* $p_\varphi(s)$ *to* $z_c(s)$ *and* $\varphi(s)$, *the vertical displacement* z_c *and pitch angle* φ *are then bounded by the arbitrarily small constants.*

(2) the ride holding constraints can be guaranteed as long as the tuning parameters and the initial values satisfy

$$(k_{ti} + b_{ti})\sqrt{\frac{q}{\lambda_{\min}(P)}} + k_{ti}\left\|z_{oi}\right\|_\infty + b_{ti}\left\|\dot{z}_{oi}\right\|_\infty \leq S_i, i = 1, 2,$$

where $q = \max\left\{V_z(0), \frac{2\varepsilon_2}{\varepsilon_1} - V_z(0)\right\}$, *and* ε_1, ε_2 *are positive constants and* $V_z(0)$ *is the initial value of the Lyapunov candidate of zero dynamics, which will be defined later.*

Proof To prove the stability of the closed-loop systems, firstly, we choose the Lyapunov function candidate as shown in (7.11), and its derivative of $V(t)$ can be given as

$$\begin{aligned}
\dot{V}(t) =& M p_z \dot{p}_z + I p_\varphi \dot{p}_\varphi + r_M^{-1}\tilde{M}\dot{\hat{M}} + r_I^{-1}\tilde{I}\dot{\hat{I}} \\
&+ \sum_{i=1}^{2}(e_{F_i}\dot{e}_{F_i} + r_i^{-1}\tilde{\beta}_{ei}\dot{\hat{\beta}}_{ei} + y_{F_i}\dot{y}_{F_i}) \\
=& - k_{pz}p_z^2 - k_{p\varphi}p_\varphi^2 + \sum_{i=1}^{2}(-k_{ei}e_{F_i}^2 + p_z y_{F_i}) \\
&+ a y_{F_1} p_\varphi - b y_{F_2} p_\varphi + \sum_{i=1}^{2}(y_{F_i}\dot{y}_{F_i}) \\
&+ \tilde{M}(r_M^{-1}\dot{\hat{M}} - k_z\dot{z}_c p_z) + \tilde{I}(r_I^{-1}\dot{\hat{I}} - k_\varphi\dot{\varphi}p_\varphi) \\
&+ \sum_{i=1}^{2}\tilde{\beta}_{ei}\left[r_i^{-1}\dot{\hat{\beta}}_{ei} - \Theta_i(\cdot)e_{Fi}\right].
\end{aligned}$$

Based on the designed adaptive laws (7.6), (7.7) and (7.9), and using Young's inequalities, we have

$$\dot{V}(t) \leq -(k_{pz} - \frac{1}{2})p_z^2 - (k_{p\varphi} - \frac{1}{2}a^2 - \frac{1}{2}b^2)p_\varphi^2 - \sum_{i=1}^{2}k_{ei}e_{F_i}^2$$

$$-\sum_{i=1}^{2}(\frac{1}{\tau_i}-2)y_{F_i}^2+\frac{1}{4}\sum_{i=1}^{2}\Delta_i^2(t).$$

If we keep the parameters to satisfy the parameter conditions $\sum_p$, then we have

$$\begin{aligned}\dot{V}(t)\leq&-\lambda Mp_z^2-\lambda Ip_\varphi^2-\lambda\sum_{i=1}^{2}e_{F_i}^2-\lambda\sum_{i=1}^{2}y_{F_i}^2\\&-\lambda r_M^{-1}\tilde{M}^2-\lambda r_I^{-1}\tilde{I}^2-\lambda\sum_{i=1}^{2}r_i^{-1}\tilde{\beta}_{ei}^2\\&+\frac{1}{4}\sum_{i=1}^{2}\Delta_i^2(t)+\lambda r_M^{-1}\tilde{M}^2+\lambda r_I^{-1}\tilde{I}^2+\lambda\sum_{i=1}^{2}r_i^{-1}\tilde{\beta}_{ei}^2\\\leq&-2\lambda V(t)+R_0,\end{aligned}\tag{7.16}$$

where $R_0=\frac{1}{4}\sum_{i=1}^{2}\Delta_i^2(t)+\lambda r_M^{-1}(M_{\max}-M_{\min})^2+\lambda r_I^{-1}(I_{\max}-I_{\min})^2+\lambda\sum_{i=1}^{2}r_i^{-1}(\beta_{ei\max}-\beta_{ei\min})^2$.

Inequality (7.16) can further result in

$$V(t)\leq\frac{R_0}{2\lambda}+(V(0)-\frac{R_0}{2\lambda})e^{-2\lambda t},\tag{7.17}$$

which implies all the signals in $V(t)$ are bounded (in this stage, the final bound can not be made arbitrarily small).

Next, we will show that all the error signals are uniformly ultimately bounded and the steady-state tracking error can be made arbitrarily small. Define a positive definite function $\bar{V}(t)$ satisfying

$$\bar{V}(t)=\frac{1}{2}Mp_z^2+\frac{1}{2}Ip_\varphi^2+\frac{1}{2}\sum_{i=1}^{2}(e_{F_i}^2+y_{F_i}^2).\tag{7.18}$$

In order to made a contradiction, we assume that there exist $T>0$ so that when $t>T$,

$$\bar{V}(t)>\frac{R}{2\lambda}+\sigma,\tag{7.19}$$

where $R=\frac{1}{4}\sum_{i=1}^{2}\Delta_i^2(t)$ and σ is a positive constant which can be set arbitrarily small. On the other hand, the following inequations is true

$$\begin{aligned}\dot{V}(t)\leq&-2\lambda V(t)+R+\lambda r_M^{-1}\tilde{M}^2+\lambda r_I^{-1}\tilde{I}^2+\lambda\sum_{i=1}^{2}r_i^{-1}\tilde{\beta}_{ei}^2\\=&-2\lambda\bar{V}(t)+R.\end{aligned}$$

Integrating both sides of the above inequality from zero to any $t > 0$, we obtain

$$V(t) - V(0) \leq \int_0^t (-2\lambda \bar{V}(t) + R)dt.$$

Because $V(t)$ is bounded, we have $f(t) = \int_0^t (2\lambda \bar{V}(t) - R)dt \leq V(0) - V(t)$ is bounded as well. It is obvious that $\ddot{f}(t) = 2\lambda \dot{\bar{V}}(t) = 2\lambda(Mp_z\dot{p}_z + Ip_\varphi \dot{p}_\varphi + \sum_{i=1}^2 (e_{F_i}\dot{e}_{F_i} + y_{F_i}\dot{y}_{F_i}))$ is bounded, and then, based on Barbalat's lemma, we have

$$\lim_{t\to\infty} \dot{f}(t) = \lim_{t\to\infty} (2\lambda \bar{V}(t) - R) = 0,$$

which implies

$$\bar{V}(t) \leq \frac{R}{2\lambda} + \sigma. \tag{7.20}$$

Note that σ can be chosen arbitrarily small, so the tracking errors p_z, p_φ, e_{F_1} and e_{F_2} are uniformly ultimately bounded and the steady-state tracking error can be made arbitrarily small by properly choosing tuning parameters.

The adaptive backstepping controller design yields to the subsystems of $\mathbf{M_S}$, $\mathbf{M_A}$, while the tire subsystem $\mathbf{M_T}$ is not considered in the controller design. Hence, we should prove the stability of zero dynamics. To find it, we set $p_z = p_\varphi = 0$. Then, we obtain:

$$F_1 = k_{s1}\Delta y_1 + b_{e1}\Delta \dot{y}_1, \ F_2 = k_{s2}\Delta y_2 + b_{e2}\Delta \dot{y}_2. \tag{7.21}$$

If we use F_1 and F_2 to replace ones in tire subsystems $\mathbf{M_T}$, we obtain the following zero dynamics:

$$\dot{x} = A_z x + Bz_o, \tag{7.22}$$

where

$$x = \begin{bmatrix} z_1 \\ \dot{z}_1 \\ z_2 \\ \dot{z}_2 \end{bmatrix}, \ A_z = \begin{bmatrix} 0 & 1 & 0 & 0 \\ -\frac{k_{t1}}{m_f} & -\frac{b_{t1}}{m_f} & 0 & 0 \\ 0 & 0 & 0 & 1 \\ 0 & 0 & -\frac{k_{t2}}{m_r} & -\frac{b_{t2}}{m_r} \end{bmatrix},$$

$$B = \begin{bmatrix} 0 & 0 & 0 & 0 \\ \frac{k_{t1}}{m_f} & \frac{b_{t1}}{m_f} & 0 & 0 \\ 0 & 0 & 0 & 0 \\ 0 & 0 & \frac{k_{t2}}{m_r} & \frac{b_{t2}}{m_r} \end{bmatrix}, \ z_o = \begin{bmatrix} z_{o1} \\ \dot{z}_{o1} \\ z_{o2} \\ \dot{z}_{o2} \end{bmatrix}.$$

Defining a positive function $V_z = x^T Px$, with $P > 0$ is a positive matrix, we have

$$\dot{V}_z = \dot{x}^T Px + x^T P\dot{x} = x^T(A_z^T P + A_z P)x + 2x^T PBz_o.$$

It is easy to verify that the matrix A_z has eigenvalues with negative real parts. Hence, we have $A_z^T P + A_z P = -Q$, where $Q > 0$ is a positive matrix. Noting that

$$2x^T PBz_o \leq \frac{1}{\upsilon_1} x^T PBB^T Px + \upsilon_1 z_o^T z_o,$$

where υ_1, υ_2 are tuning positive values, the following inequality is obtained.

$$\begin{aligned}\dot{V}_z \leq& - x^T Qx + \frac{1}{\upsilon_1} x^T PBB^T Px + \upsilon_1 z_o^T z_o \\ \leq&[-\lambda_{\min}(P^{-\frac{1}{2}} QP^{-\frac{1}{2}}) + \frac{1}{\upsilon_1}\lambda_{\max}(P^{\frac{1}{2}} BB^T P^{\frac{1}{2}})]V_z + \upsilon_1 z_o^T z_o.\end{aligned}$$

Based on properly chosen matrices P, Q and tuning values υ_1, we can find a positive value ε_1, so that

$$\dot{V}_z \leq -\varepsilon_1 V_z + \varepsilon_2, \tag{7.23}$$

where $\varepsilon_2 = \upsilon_1 z_{o\max}$ with $z_o^T z_o \leq z_{o\max}$. Equation (7.23) shows that the Lyapunov function is bounded by

$$V_z(t) \leq (V_z(0) - \frac{\varepsilon_2}{\varepsilon_1})e^{-\varepsilon_1 t} + \frac{\varepsilon_2}{\varepsilon_1}, \tag{7.24}$$

which tells us that $|z_i| \leq \sqrt{\frac{q}{\lambda_{\min}(P)}}$, $|\dot{z}_i| \leq \sqrt{\frac{q}{\lambda_{\min}(P)}}$ $i = 1, 2$, with

$$q = \max\left\{V_z(0), \frac{2\varepsilon_2}{\varepsilon_1} - V_z(0)\right\}.$$

Then, we can estimate the ride holding performance as

$$\begin{aligned}&|k_{ti}(z_i - z_{oi}) + b_{ti}(\dot{z}_i - \dot{z}_{oi})| \\ \leq&(k_{ti} + b_{ti})\sqrt{\frac{q}{\lambda_{\min}(P)}} + k_{ti}\,\|z_{oi}\|_\infty + b_{ti}\,\|\dot{z}_{oi}\|_\infty.\end{aligned} \tag{7.25}$$

which implies that if we adjust the tuning parameters and initial values to guarantee

$$(k_{ti} + b_{ti})\sqrt{\frac{q}{\lambda_{\min}(P)}} + k_{ti}\,\|z_{oi}\|_\infty + b_{ti}\,\|\dot{z}_{oi}\|_\infty \leq S_i, i = 1, 2,$$

then we can keep the ratio of tire dynamic load and static load less than 1, which further guarantee the ride safety. The proof of this theorem is completed.

Remark 7.4 In the proposed method, we can see that the satisfactory closed-loop stability with suitable transient performance can be achieved by properly adjusting design parameters k_z, k_{pz}, k_ϕ, $k_{p\phi}$, k_{e1} and k_{e2}. Actually, in a viewpoint of closed-loop system bandwidth, higher control gains bring about higher bandwidth of the closed-loop system, which results in better tracking performance. However, high bandwidth of the closed-loop system means that there will introduce high frequency disturbances, which may degrade the system performance. Thus, in the process of implementation, we may turn the gain parameters from minor until the desired trajectory performance is obtained.

Remark 7.5 Without consideration of cost and complexity, all the signals can definitely be measured, which provides the basis for this approach. However, although the measurements of all signals can be realized, this process, sometimes, can be added measurement noises. In this way, it is better to use filtering process to reduce measurement errors and noises. As a main research target in the future, it is essential that we should consider the robust control strategy for the measured signals with measurement noises for the active suspension systems.

7.1.3 Simulation Verification

In this section, we provide an example to illustrate the effectiveness of the proposed approach. The half-car model parameters are listed in Table 7.1. Give the initial conditions as: $\hat{M}(0) = 1350$, $\hat{I}(0) = 650$, $\hat{\beta}_e(0) = 800$ Mpa and the rest ones are assumed as zeros. The controller parameters are given in Table 7.2.

In the following, we will illustrate the performances of the closed-loop suspension system. For suspension control systems, evaluation of the suspension performance is based on the examination of several response quantities, that is, the

Table 7.1 The model parameters of half-car active suspensions

Parameter	Value	Parameter	Value
M	1200 kg	I	600 kgm^2
$M_{\max}$	1500 kg	$I_{\max}$	700 kgm^2
$M_{\min}$	1000 kg	$I_{\min}$	500 kgm^2
$m_f = m_r$	100 kg	$k_{nf1} = k_{nr1}$	1000 N/m
$b_{e1} = b_{e2}$	2500 Ns/m	$b_{t1} = b_{t2}$	1000 N/m
$k_{s1} = k_{s2}$	15000 N/m	a	1.2 m
$k_{t1} = k_{t2}$	200000 N/m	b	1.5 m
P_s	5×10^6 Pa	A	0.006 m^2
μ	2.32×10^8	β_{ei} $(i = 1, 2)$	700 Mpa
C_{tp}	4×10^{-13}	$\beta_{ei\max}$	800 Mpa
V_t	1.2×10^{-3} m^3	$\beta_{ei\min}$	200 Mpa
V_0	20 m/s		

Table 7.2 The controller parameters of active suspensions

Parameter	Value	Parameter	Value
$k_{e1} = k_{e2}$	1000	$r_M = r_I$	0.01
$k_z = k_p$	200	$r_{\theta_1} = r_{\theta_2}$	0.01
$k_{pz} = k_{p\varphi}$	200	$\tau_1 = \tau_2$	0.001

vertical displacement of car body $z_c(t)$, the pitch displacement of car body $\varphi(t)$ and their corresponding accelerations $\ddot{z}_c(t)$, $\ddot{\varphi}(t)$. Moreover, the tire dynamic load $k_{ti}(z_i - z_{oi}) + b_{ti}(\dot{z}_i - \dot{z}_{oi})$ should be less than the static load S_i, $i = 1, 2$. And our final target is to design a control law such that: (1) the vertical displacement of car body $z_c(t)$, the pitch displacement of car body $\varphi(t)$ and their corresponding accelerations $\ddot{z}_c(t)$, $\ddot{\varphi}(t)$ can be as small as possible; (2) the ride safety constraint should be guaranteed, that is $\left|\frac{k_{ti}(z_i - z_{oi}) + b_{ti}(\dot{z}_i - \dot{z}_{oi})}{S_i}\right| < 1$.

In order to evaluate the suspension characteristics with respect to ride comfort and safety, the variability of the road profiles is taken into account. The proposed controllers are tested by the classic bump road inputs. Bump road inputs can be generally assumed as shocks which are discrete events of relatively short duration and high intensity, caused by, for example, a pronounced bump or pothole on an otherwise smooth road. The corresponding ground displacement is given by

$$z_{o1} = \begin{cases} \frac{h_b[1-\cos(6\pi t)]}{2}, & 1 \le t \le 1.25, \\ 0, & \text{otherwise}, \end{cases} \tag{7.26}$$

where h_b is the height of the bump road input. It is assumed that the bump road input has the magnitude for $h_b = 2\,\text{cm}$, and road conditions for the front and rear wheels are the same but with a time delay of $(a + b)/V_0$, where V_0 is the velocity of the vehicle. For subsequent comparison, the following four systems are given respectively:

S1: Passive suspension;

S2: Active suspension systems with standard adaptive backstepping control strategy (without consideration of electro-hydraulic actuators dynamics);

S3: Active suspension systems with the proposed controller.

Figure 7.2 shows the responses of vertical displacements of car body, and for comparison, both active control and passive systems are plotted in this figure. We

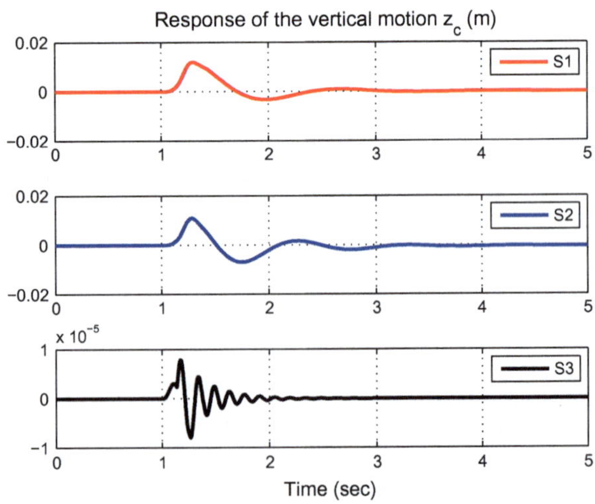

Fig. 7.2 The vertical displacements of car body

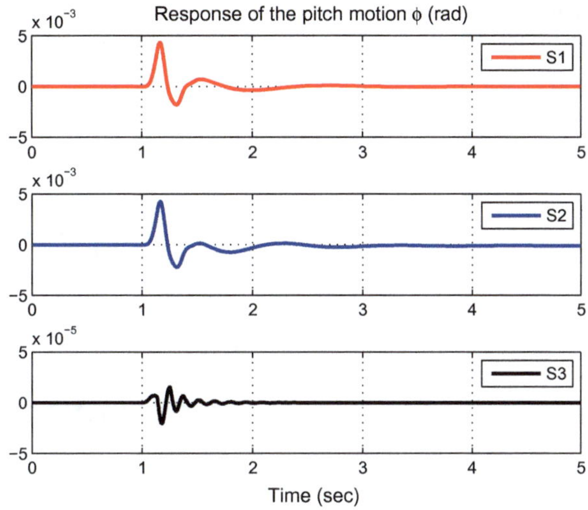

Fig. 7.3 The pitch angle displacement of car body

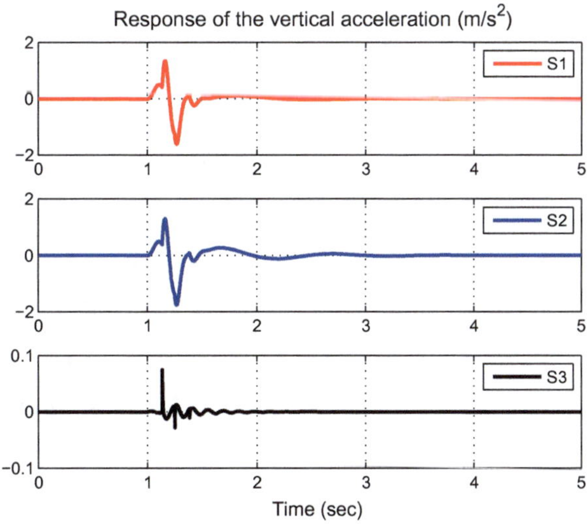

Fig. 7.4 The dynamic load of the front wheel

can see that both peak and steady time have been significantly improved in active systems with our proposed controllers.

The time histories of pitch motion of car body are plotted in Fig. 7.3, where we can see that the impact from roughness road on the angle φ is very small compared with other systems. This implies that our designed controllers can stabilize the attitude substantially.

It is well known that vehicle body accelerations are strictly related to ride comfort of passengers, and thus, the body accelerations (both vertical and pitch accelerations) for active and passive suspensions should be given and compared to illustrate the

Fig. 7.5 The dynamic load of the rear wheel

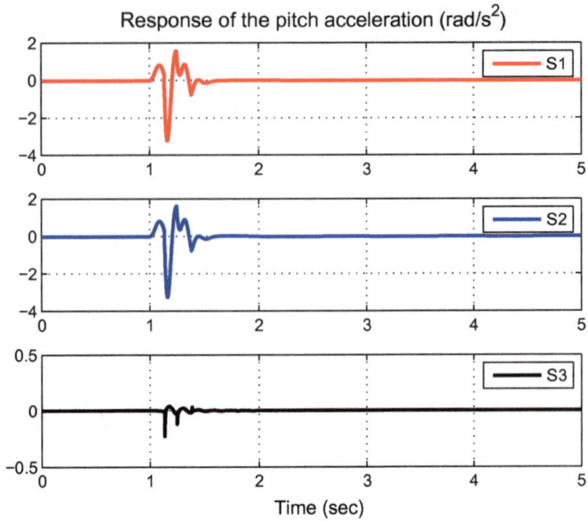

Fig. 7.6 The vertical accelerations of car body

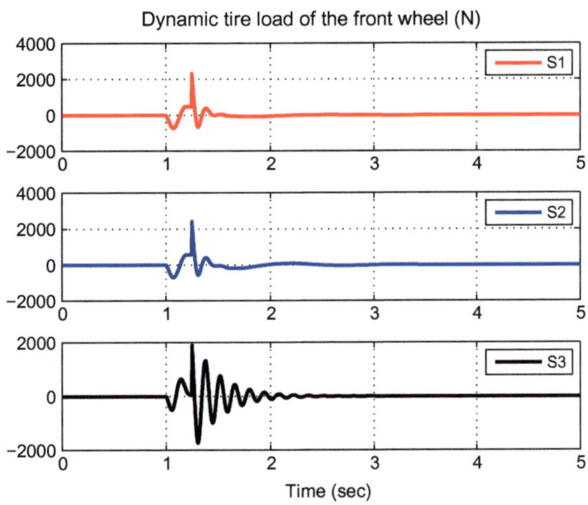

effectiveness of our controllers. In Figs. 7.6 and 7.7, the curves of acceleration are plotted, where Fig. 7.6 stands for the history of vertical motion, and Fig. 7.7 represents the history of pitch motion. It can be seen that active suspensions can achieve smaller peaks of accelerations, which contributes a better ride comfort of passengers.

Remark 7.6 In the procedure of designing control law, although we choose the vertical displacement as the objective trajectory, yet from the simulation results, we can see that the acceleration signals can be also handled well, whose precision has reached the 10^{-1}. Actually, vertical displacement has been selected as a tracking index in many references, such as [9, 10], whose problems are similar to ours. This shows the reasonability of this selection from indirect sources.

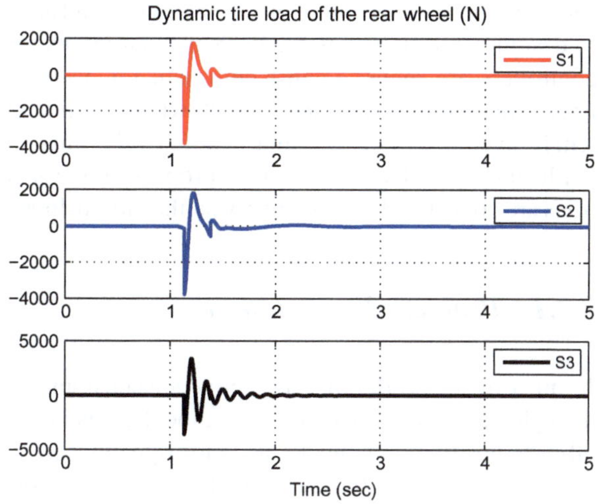

Fig. 7.7 The pitch angle accelerations of car body

The static tire loads for both front and rear wheels can be calculated by (7.5) as $S_1 = 7.5133 \times 10^3$ N, $S_2 = 6.2067 \times 10^3$ N. Figures 7.4 and 7.5 show the responses of dynamic tire load of the two wheels, and the peaks of dynamic tire load for the two wheels are all within the bounds.

7.1.4 Conclusion

In this section, an adaptive vibration control strategy has been proposed for the nonlinear uncertain suspension systems to stabilize both the vertical and pitch motions of the car. Simultaneously, the ride holding performances have been preserved within their allowable limits in the controller design. Moreover, this section considered the electro-hydraulic systems as actuators to supply the active forces into suspension systems, and to overcome the "exploration of terms" problem existing in standard backstepping, a filter-based backstepping control strategy was subsequently employed. Finally, a design example has been shown to illustrate the effectiveness of the proposed active controllers.

7.2 Adaptive Robust Vibration Control of Full-Car Active Suspensions with Electro-Hydraulic Actuators

In this section, a full-car model is adopted and electro-hydraulic actuators with highly nonlinear characteristics are considered to form the basis of accurate control. In this study, the H_∞ performance is introduced to realize the disturbance suppression by selecting the actuator forces as virtual inputs, and an ARC technology is further used to design controllers which help real force inputs track virtual ones. The resulting controllers are robust against both actuator parametric uncertainties and actuator uncertain nonlinearities, and the following stability analysis for the closed-loop system is given within the Lyapunov framework. Finally, a numerical example is shown to illustrate the effectiveness of the proposed control law, where different road conditions are considered in order to reveal the closed-loop system performance in details.

7.2.1 Problem Formulation

In most of the works treating active suspensions, the effect of actuator dynamics is neglected and the actuators are modelled as an ideal force generator. Additionally, studies that do consider actuator dynamics in suspension design are based on the assumption that the actuator dynamics are known exactly without parametric uncertainties and unmodelled nonlinearities. Such assumption results in inaccuracies of controller design in actual engineering. In active suspensions design, electro-hydraulic systems could be a good alternative as the actuators to isolate the vibrations transmitted to the passengers, because they are more powerful and less bulky compared with other actuators. However, the main drawback of electro-hydraulic systems is their highly non-linear behavior, making it more difficult to design the ideal control law. Furthermore, aside from the nonlinear nature of hydraulic dynamics, hydraulic servo systems are also subject to excessive model uncertainties, and many works research the uncertain systems [5, 11]. In actual implementations, interactions between hydraulic actuators and vehicle dynamics make it difficult to track forces, especially in the presence of uncertainties. It is worthwhile to mention that, in [12], the authors proposed a two-loop design approach where the control architecture contained a force-loop controller and a main-loop controller. In main-loop controller design, the LQG approach was used to achieve the performance of active suspensions, whereas in force-loop controller design, the adaptive robust controller was proposed to track the desired force commanded by the main-loop controller accurately. The proposed method took effects in applications, but the stability analysis of combined system including two loops was not given. Based on the above discussions, for active suspensions with the nonlinear uncertain actuators, it is necessary to develop a systematical control scheme such that the closed-loop stability is guaranteed.

In this study, the problem of vibration suppression is investigated to improve the ride comfort of the active suspension systems, where the full-car active suspension systems with electro-hydraulic actuators are considered to cover a wider applications.

In this section, a full-car electro-hydraulic suspension model is considered , which has seven DOF due to the heave, pitch and roll motions of the sprung mass and the vertical motions of the unsprung masses, as depicted in Fig. 1.4, where definitions of the variables are stated in details. The dynamics of the electro-hydraulic actuators are taken into consideration as shown in Fig. 7.1. In the study, the valve dynamics are neglected and the servo-valve opening x_{vi} is directly related to the control input by a known static mapping. Neglecting the effect of leakage flows in the cylinder, the cylinder dynamics can be written as follows

$$\dot{P}_{Li} = -\beta P_{Li} - \alpha A \Delta \dot{y}_i + \mu x_{vi} g_i(\cdot),$$

where $P_{Li} = P_{1i} - P_{2i}$ is the load pressure of the cylinder, A is the ram area of the cylinder, $\alpha = \frac{4\beta_e}{V_t}$, $\beta = \alpha C_{tp}$, $\mu = \alpha C_d \omega \sqrt{\frac{1}{\rho}}$, $g_i(\cdot) = \sqrt{P_s - \text{sgn}(x_{vi}) P_{Li}}$, β_e is the effective bulk modulus, V_t is the total actuator volume, C_{tp} is the total leakage coefficient of the piston, ρ is the hydraulic fluid density, ω is the spool valve area gradient, C_d is the discharge coefficient and P_s is supply pressure. Noting that the forces generated by actuators can be described as $f_i = AP_{Li}$, the force dynamics are given as

$$\dot{f}_i = -\beta f_i - \alpha A^2 \Delta \dot{y}_i + A \mu x_{vi} g_i(\cdot). \tag{7.27}$$

Here, the open-loop natural frequency of the actuation systems can be estimated as $\omega_n = \sqrt{\frac{4\beta_e}{V_t}\frac{A^2}{M}}$. By selecting the parameters of the electro-hydraulic systems as $\beta_e =$ 700 Mpa (200 − 800 Mpa), $V_t = 1.2 \times 10^{-3}\,\text{m}^3$, we have $f = \omega_n / 2\pi = 42.1$Hz. As we know, the sensitive frequency of the active suspensions is usually 4 − 8 Hz, which implies that the electro-hydraulic systems can be used as the actuator for the vehicle suspension control without worrying about the dynamics of the hydraulic actuation system, and the designer can easily find a controller to make the closed-loop response fast enough to meet the requirement of the vehicle suspension control in the concerned range.

Considering the hydraulic fluid's bulk modulus β_e as the uncertain parameter and defining actuator force $F = \begin{bmatrix} f_1 & f_2 & f_3 & f_4 \end{bmatrix}^T$, we can obtain the following dynamics

$$\dot{F} = \theta_1(-C_{tp} F - A^2 \dot{z}_2 + \eta) + d_0, \tag{7.28}$$

where $\theta_1 = \alpha = \frac{4\beta_e}{V_t}$ is an unknown parameter, d_0 denotes general uncertainties arise from unmodeled dynamics, $\eta = \begin{bmatrix} \eta_1 & \eta_2 & \eta_3 & \eta_4 \end{bmatrix}^T$ and $\eta_i = AC_d \omega \sqrt{\frac{1}{\rho}} x_{vi} g_i(\cdot)$. θ_1 and d_0 are assumed to be bounded and their bounds are known as $0 < \theta_{1m} < \theta_1 < \theta_{1M}$ and $\|d_0\| < d_M$. z_2 respects the suspension spaces which will be defined later.

The dynamic equations of motion for the full vehicle model (1.7) with electro-hydraulic actuators are obtained in the Sect. 1.2, based on an assumption that the pitch and roll angles are small.

7.2.2 ARC-Based H_∞ Control Law Synthesis

Step 1: Design a desired virtual force $F_a(t)$, such that if $F(t) = F_a(t)$, then the L_2 gain from disturbances to the output z_1 is smaller or less than a certain given value γ.

Defining the state variable $x = \left[z\ \theta\ \varphi\ y_1\ y_2\ y_3\ y_4\ \dot{y}\ \dot{\theta}\ \dot{\varphi}\ \dot{y}_1\ \dot{y}_2\ \dot{y}_3\ \dot{y}_4\right]^T$, output variables $z_1 = [z\ \theta\ \varphi]^T$, $z_2 = [\Delta y_1\ \Delta y_2\ \Delta y_3\ \Delta y_4]^T$, and disturbance $w = [y_{o1}\ y_{o2}\ y_{o3}\ y_{o4}]^T$, the state-space form of car body system Σ_p can be expressed as

$$\Sigma_p \begin{cases} \dot{x}(t) = A_p x(t) + B_w w(t) + B_p F(t), \\ z_1(t) = C_1 x(t), \\ z_2(t) = C_2 x(t), \end{cases} \tag{7.29}$$

where

$$A_p = \begin{bmatrix} 0_{7\times7} & I_7 \\ E^{-1}G_1 & E^{-1}G_2 \end{bmatrix},\ B_w = \begin{bmatrix} 0_{7\times4} \\ E^{-1}F_1 \end{bmatrix},\ B_p = \begin{bmatrix} 0_{7\times4} \\ E^{-1}F_2 \end{bmatrix},$$
$$E = \text{diag}\left\{ M\ I_z\ I_x\ m_1\ m_2\ m_3\ m_4 \right\},\ B = \text{diag}\left\{ b_1\ b_2\ b_3\ b_4 \right\},$$
$$K_t = \text{diag}\left\{ k_{t1}\ k_{t2}\ k_{t3}\ k_{t4} \right\},\ K = \text{diag}\left\{ k_1\ k_2\ k_3\ k_4 \right\},\ C_1 = \left[I_3\ 0_{3\times11} \right],$$
$$G_1 = -F_2KT - \left[0_{7\times3}\ F_1 \right],\ G_2 = -F_2BT,\ T = \left[L\ -I_4 \right],\ C_2 = \left[T\ 0_{4\times7} \right].$$
$$L = \begin{bmatrix} 1 & 1 & 1 & 1 \\ a & a & -b & -b \\ -c & d & -c & d \end{bmatrix}^T,\ F_1 = \begin{bmatrix} 0_{3\times4} \\ K_t \end{bmatrix},\ F_2 = \begin{bmatrix} L^T \\ -I_4 \end{bmatrix}.$$

The output z_1 in (7.29) is defined to describe vertical, pitch and roll motions of the vehicle, and our goal is to stabilize the output z_1, keep it at minimum, in order to ensure passengers' comfort. The suspension spaces are expressed as output z_2, each component in which should not exceed the allowable maximum $y_{i\,\max}$, that is, $|\Delta y_i| \leq y_{i\,\max}$.

It is our expectation to synthesize a desired control function $F_a(t)$ for the virtual control $F(t)$, so that the expected performances can be matched. However, $F(t)$ is not a real input, but a state. Here, by denoting $e_1(t) = F(t) - F_a(t)$ as the error between $F(t)$ and $F_a(t)$, we obtain

$$\tilde{\Sigma}_p \begin{cases} \dot{x}(t) = A_p x(t) + B_p F_a(t) + B_p e_1(t) + \Delta(t), \\ z_1(t) = C_1 x(t), \\ z_2(t) = C_2 x(t), \end{cases} \tag{7.30}$$

where $\Delta(t) = B_w w(t)$ is the lumped disturbances. The following lemma gives the solution to design the expected control function $F_a(t) = Hx(t)$.

Lemma 7.7 *Consider the car body system in (7.30) with $e_1(t) = 0$. If there exist positive scalars γ, σ and symmetric matrices H, $P > 0$ satisfying*

$$\begin{bmatrix} A_p^T P + H^T B_p^T P + P A_p + P B_p H & P & C_1^T \\ * & -\gamma^2 I & 0 \\ * & * & -I \end{bmatrix} < 0, \tag{7.31}$$

$$\begin{bmatrix} -I & \sqrt{\sigma}\{C_2\}_i \\ * & -y_{i\,\max}^2 P \end{bmatrix} < 0, \tag{7.32}$$

and the lumped disturbance energy satisfying $\|\Delta\|_2 \leq \sigma/\gamma^2$, *then a virtual input* $F_a(t)$ *exists in the form of* $F_a(t) = Hx(t)$ (H *is a constant matrix*), *such that*

1. *under zero initial conditions,* $\|z_1\|_2 < \gamma \|\Delta\|_2$;
2. *suspension spaces travel within their allowed ranges.*

Proof This proof is omitted, and one can obtain this conclusion based on [13].

To obtain the solvable form of controller gain H, pre- and post-multiplying (7.31)–(7.32) by $J_1 = \text{diag}\{P^{-1},\ I,\ I\}$, $J_2 = \text{diag}\{I,\ P^{-1}\}$ and their transposes, respectively, together with the definitions of $\bar{P} = P^{-1}$, $\bar{H} = HP^{-1}$, we can obtain the solvable forms:

$$\begin{bmatrix} \bar{P}A_p^T + \bar{H}B_p^T + A_p\bar{P} + B_p\bar{H} & I & \bar{P}C_1^T \\ * & -\gamma^2 I & 0 \\ * & * & -I \end{bmatrix} < 0, \tag{7.33}$$

$$\begin{bmatrix} -I & \sqrt{\sigma}\{C_2\}_i\bar{P} \\ * & -y_{i\max}^2\bar{P} \end{bmatrix} < 0. \tag{7.34}$$

If inequalities (7.33)–(7.34) have a feasible solution, the matrix H is given by $H = \bar{H}\bar{P}^{-1}$.

Step 2: Synthesize an actual control law for $u = x_v$, so that $F(t)$ tracks the desired control function $F_a(t)$ in the presence of unknown parameter θ_1 and general uncertainties d_0.

Differentiating the error dynamics for $e_1 = F - F_a$, we have

$$\dot{e}_1 = \dot{F} - \dot{F}_a = \theta_1(-C_{tp}F - A^2\dot{z}_2 + \eta) + d_0 - \dot{F}_a. \tag{7.35}$$

Design an ARC control law as $\eta = \eta_a + \eta_s$, where η_a is used to achieve an improved adaptive model compensation and η_s functions as the robust term. The adaptive part η_a is designed as:

$$\eta_a = C_{tp}F + A^2C_2(A_px + B_pF) + \frac{1}{\hat{\theta}_1}(H(A_px + B_pF) - 2B_p^TPx - \varepsilon_1 e_1), \tag{7.36}$$

where $\hat{\theta}_1$ is a bounded projected estimate of θ_1 and ε_1 is used for tuning the controller. The adaptation law is chosen as the projection type with the following form:

$$\dot{\hat{\theta}}_1 = Proj_{\hat{\theta}_1}(r\tau) = \begin{cases} 0, \text{ if } \hat{\theta}_1 = \theta_{1M} \text{ and } r\tau > 0, \\ 0, \text{ if } \hat{\theta}_1 = \theta_{1m} \text{ and } r\tau < 0, \\ r\tau, \text{ otherwise.} \end{cases}, \tag{7.37}$$

where $r > 0$ is a tunable gain and $\tau = e_1^T(-C_{tp}F - A^2C_2(A_px + B_pF) + \eta)$.

The robust control function η_s is now chosen to satisfy the following conditions:

$$\begin{aligned}&\text{condition 1:}\quad e_1^T[\hat{\theta}_1\eta_s - \tilde{\theta}_1(-C_{tp}F - A^2C_2(A_px + B_pF) + \eta) + \bar{d}] \le \varepsilon_2,\\ &\text{condition 2:}\quad e_1^T\hat{\theta}_1\eta_s \le 0,\end{aligned} \tag{7.38}$$

where $\tilde{\theta}_1 = \hat{\theta}_1 - \theta_1$, $\bar{d} = d_0 - (\theta_1 A^2 C_2 + H)\Delta$ and ε_2 is a design parameter which can be arbitrarily small. Let h be any smooth function satisfying

$$h \ge \frac{1}{\varepsilon_{2a}}(\theta_{1M} - \theta_{1m})^2 \left\| -C_{tp}F - A^2C_2(A_px + B_pF) + \eta_a \right\|^2 + \frac{1}{\varepsilon_{2b}}\bar{d}_M^2,$$

and $\bar{d}_M$ is a positive number such that $\|\bar{d}\| \le \bar{d}_M$, ε_{2a} and ε_{2b} are adjustable small positive numbers satisfying $\varepsilon_{2a} + \varepsilon_{2b} = \varepsilon_2$. Then, based on [14], the robust control part η_s can be chosen as:

$$\eta_s = -\frac{1}{4\theta_{1m}}he_1. \tag{7.39}$$

Finally, the original control signal $u_i = x_{vi}$ can then be found from the definition of η, i.e.

$$u_i = \frac{\eta_i}{AC_d\omega\sqrt{\frac{1}{\rho}}\sqrt{P_s - \text{sgn}(\eta_i)P_{Li}}}, \quad i = 1, 2, 3, 4. \tag{7.40}$$

Theorem 7.8 *With the ARC law (7.40) and the projection type adaptive law (7.37), the following results hold:*

A. In general (i.e., the system is subjected to parametric uncertainties, unmodelled uncertainties and external disturbances), all signals in system Σ_I are bounded;

B. If after a finite time, the system is subjected to parametric uncertainties only, then the state x and the tracking error e_1 asymptotically converge to zero.

Proof of Part A: Choose a positive definite function as $V = x^T Px + \frac{1}{2}e_1^T e_1$, and we have

$$\dot{V} = x^T(A_p^T P + H^T B_p^T P + PA_p + PB_pH)x + 2x^T P\Delta + 2e_1^T B_p^T Px + e_1^T\dot{e}_1. \tag{7.41}$$

Substituting the ARC law (7.40) into (7.41) results in

$$\dot{V}(t) \le x^T(A_p^T P + H^T B_p^T P + PA_p + PB_pH)x + 2x^T P\Delta - \varepsilon_1 e_1^T e_1 + \varepsilon_2. \tag{7.42}$$

From (7.31), we have $A_p^T P + H^T B_p^T P + PA_p + PB_pH < -\gamma^{-2}PP - C_1^T C_1$, and noting $2x^T P\Delta \le \frac{1}{\lambda_1}x^T PPx + \lambda_1\Delta^T\Delta$, where λ_1 is a positive value, we have

$$\dot{V}(t) < -(\gamma^{-2} - \lambda_1^{-1})x^T PPx - \varepsilon_1 e_1^T e_1 + \lambda_1\Delta^T\Delta + \varepsilon_2. \tag{7.43}$$

Choosing λ_1 so that $\gamma^{-2} - \lambda_1^{-1} > 0$, and defining $\epsilon_0 = \min\{\lambda_{\min}(P)(\gamma^{-2} - \lambda_1^{-1}), 2\varepsilon_1\}$, where $\lambda_{\min}(P)$ is the minimal eigenvalue, we have

$$\dot{V}(t) < -\epsilon_0 V + \varepsilon_3, \tag{7.44}$$

where $\varepsilon_3 = \varepsilon_2 + \lambda_1 \|\Delta\|_\infty$, with $\Delta^T \Delta \leq \|\Delta\|_\infty$. Equation (7.44) shows that the Lyapunov function is bounded by

$$V(t) < V(0)e^{-\epsilon_0 t} - \frac{\varepsilon_3}{\epsilon_0}(1 - e^{-\epsilon_0 t}), \tag{7.45}$$

which implies that the states and force tracking error (e_1) are bounded.

Proof of B: Choose a positive function as $V = x^T P x + \frac{1}{2} e_1^T e_1 + \frac{1}{2r} \tilde{\theta}_1^2$, and we have

$$\begin{aligned} \dot{V}(t) &= x^T(A_p^T P + H^T B_p^T P + P A_p + P B_p H)x + 2e_1^T B_p^T P x + e_1^T \dot{e}_1 + r^{-1}\tilde{\theta}_1 \dot{\hat{\theta}}_1 \\ &< -\gamma^{-2} x^T P P x + e_1^T(2B_p^T P x + \dot{e}_1) + r^{-1}\tilde{\theta}_1 \dot{\hat{\theta}}_1 \\ &< -\gamma^{-2} x^T P P x - \varepsilon_1 e_1^T e_1 + \tilde{\theta}_1 (r^{-1}\dot{\hat{\theta}}_1 - \tau) \end{aligned}$$

Noticing the property of the projection mapping $Proj_{\hat{\theta}_1}(r\tau)$: $\tilde{\theta}_1(r^{-1} Proj_{\hat{\theta}_1}(r\tau) - \tau) \leq 0$, we have $\dot{V} < -\gamma^{-2} x^T P P x - \varepsilon_1 e_1^T e_1 < 0$, which implies that the state x and the tracking error e_1 asymptotically converge to zero.

7.2.3 Simulation Verification

This section presents simulation results from the application of the control algorithms developed in Sect. 6.2.2. For comparison, the backstepping control technology introduced in [15] is also employed to stand out the effectiveness of our proposed method, and we can obtain the backstepping controller: $\eta = \beta F + (\alpha A^2 C_2 + H)(A_p x + B_p F) - M^T M e_1 - \Pi e_1$, where Π is a feedback gain matrix, and M is chosen based on the equation $(\alpha A^2 C_2 + H) B_w = 2M^T N$.

The model parameters borrowed from [16] are given as: $M = 1200\,\text{kg}$, $m_1 = m_2 = 25\,\text{kg}$, $m_3 = m_4 = 45\,\text{kg}$, $I_x = 550\,\text{kgm}^2$, $I_y = 1848\,\text{kgm}^2$, $k_1 = k_2 = 15000\,\text{N/m}$, $k_3 = k_4 = 17000\,\text{N/m}$, $b_i = 1500\,\text{Ns/m}$, $k_{ti} = 250000\,\text{N/m}$, $a = 1.2\,\text{m}$, $b = 1.4\,\text{m}$, $c = 0.7\,\text{m}$, $d = 0.8\,\text{m}$, $y_{i\max} = 0.1\,\text{m}$, $V = 20\,\text{m/s}$, $\mu = 2.32 \times 10^8$ $\text{N/(m}^{5/2}\,\text{kg}^{1/2})$, $C_{tp} = 4 \times 10^{-13}$, $A = 0.006\text{m}^2$, $P_s = 5 \times 10^6\,\text{Pa}$, $\alpha = 2.3 \times 10^{12}$ N/m^5. Here we assume that $\bar{d} = (-1)^{round(t)}$, which implies that we can choose $\bar{d}_M = 1$. The controller parameters are given as: $r = 10$, $\sigma = 1$, $\varepsilon_1 = 100$, $h = 1000$, $\theta_{1M} = 2.68 \times 10^{12}$ and $\theta_{1m} = 0.67 \times 10^{12}$.

In order to illustrate the effectiveness of H_∞ technology, the transfer function matrix from the disturbances w to the outputs (z, θ, ϕ) is given under the assumption

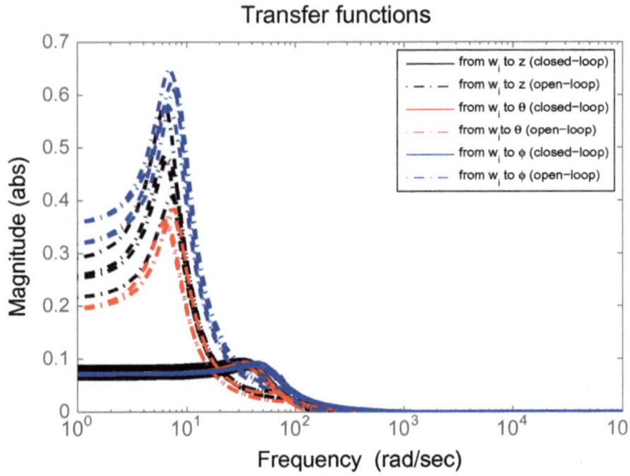

Fig. 7.8 The transfer functions from the disturbances w to the outputs (z, θ, ϕ)

of accurate force tracking, that is $G_{tf}(s) = C_1(sI - A_p - B_pH)^{-1}B_w$, and then the corresponding frequency responses are provided in Fig. 7.8, where we can see that the performances of closed-loop have been improved significantly, compared with the open-loop performances.

Bump Road Inputs

Bump road inputs can be generally assumed as shocks, and the corresponding ground displacement is given as:

$$y_o = \begin{cases} \frac{h_2[1-\cos(8\pi t)]}{2}, & 1 \leq t \leq 1.25, \\ 0, & \text{otherwise,} \end{cases} \tag{7.46}$$

where h_2 is the height of the bump road input. To validate the vibration depression effect of the designed controller, it is assumed that the bump road input has different magnitudes for the left and right wheels; 3 cm for the right and 4 cm for the left side, and road conditions for the front and rear wheels are the same but with a time delay of $(a + b)/V$.

Figure 7.9 shows comparisons among passive systems (dotted lines), active suspensions with backstepping-based H_∞ controllers (dash-dotted lines) and active suspensions with ARC-based H_∞ controllers (solid lines) for the time responses of the heave motion z, pitch motion θ and roll motion ϕ with bump road inputs in (7.46), and it is seen from these figures that the magnitudes for the heave, pitch and roll motions are significantly decreased for active suspensions with ARC-based H_∞ controllers. The reasons that ARC-based H_∞ controllers have an advantage over backstepping-based H_∞ controllers are twofold, adaptive ability (to deal with the uncertain parameter θ_1) and robust ability (to deal with the uncertain nonlinearity $\bar{d}$).

In the active suspension control, the limitations of the suspension spaces should be taken into account, which means the suspension working space must be preserved. It

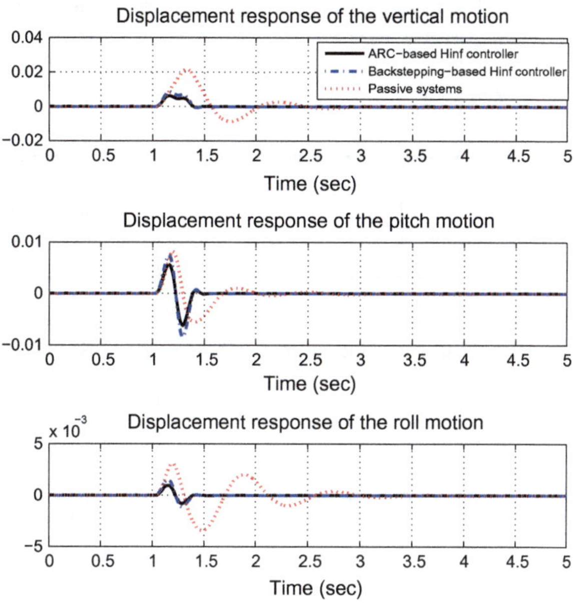

Fig. 7.9 Displacement responses for heave, pitch and roll motions

can been observed from Fig. 7.10 that the suspension space constraints of the designed systems have been guaranteed, and are all below the limitations $y_{i\max} = 0.1$m.

From the point of view of car safety, the firm uninterrupted contact of wheels to road should be ensured. The performance of road holding is considered here based on an idea constraint for road handling, that is, the dynamic tire load should not exceed the static ones for both of the front and rear wheels, i.e.

$$\frac{\sum_{i=1}^{4} k_{ti}(y_i - y_{oi})}{(M + m_1 + m_2 + m_3 + m_4)g} \leq 1.$$

From Fig. 7.11, we can see that the ratios of dynamic tire loads and static tire load are always far less than 1, which implies that the firm uninterrupted contact of wheels to road is guaranteed and the performance of ride holding is achieved when the car is running on the road.

Simulation results in Fig. 7.12 show the control inputs. Additionally, the uncertainties and unmodeled dynamics of the suspension components themselves are worthy of consideration. The simulations for the case of body uncertainties are performed Fig. 7.13, from which we can see that our designed controllers still have the capacity to suppress the disturbances in despite that there exist parameter uncertainties in suspension components themselves.

Random Road Inputs

When the road disturbances are considered as vibrations, they are consistent and typically specified as random process. Consider the road inputs as a sequence of

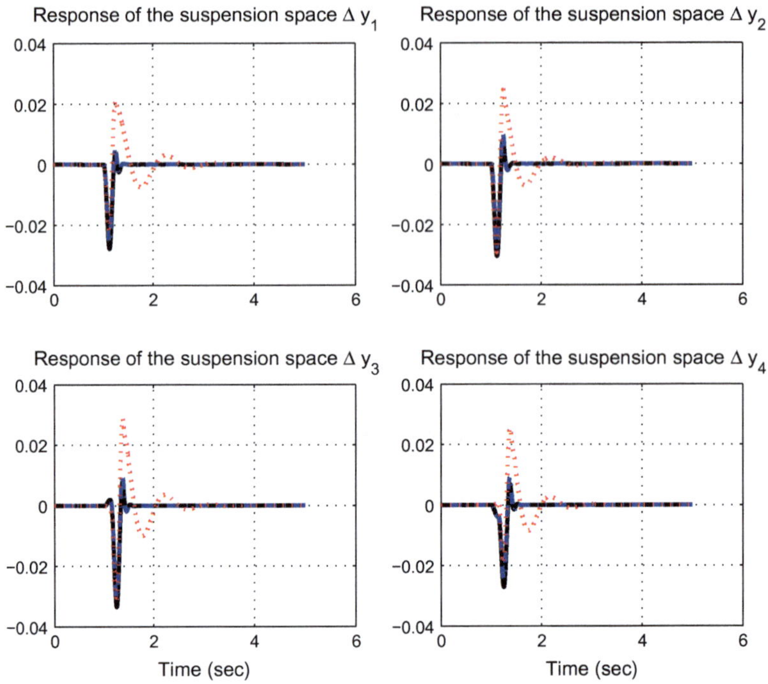

Fig. 7.10 Responses of the suspension spaces

Fig. 7.11 The ratio of dynamic tire load and static tire load

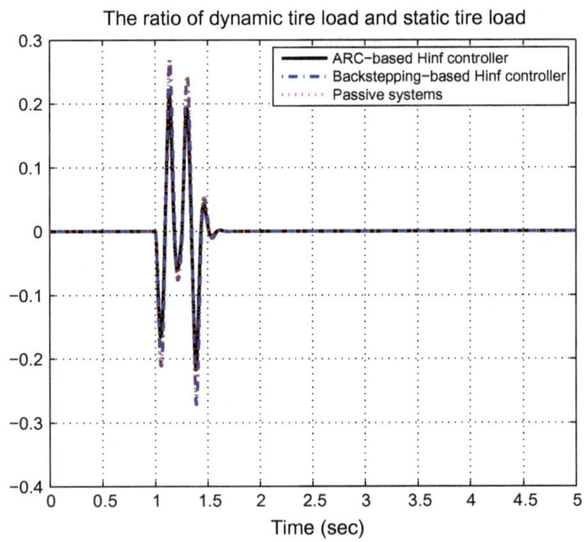

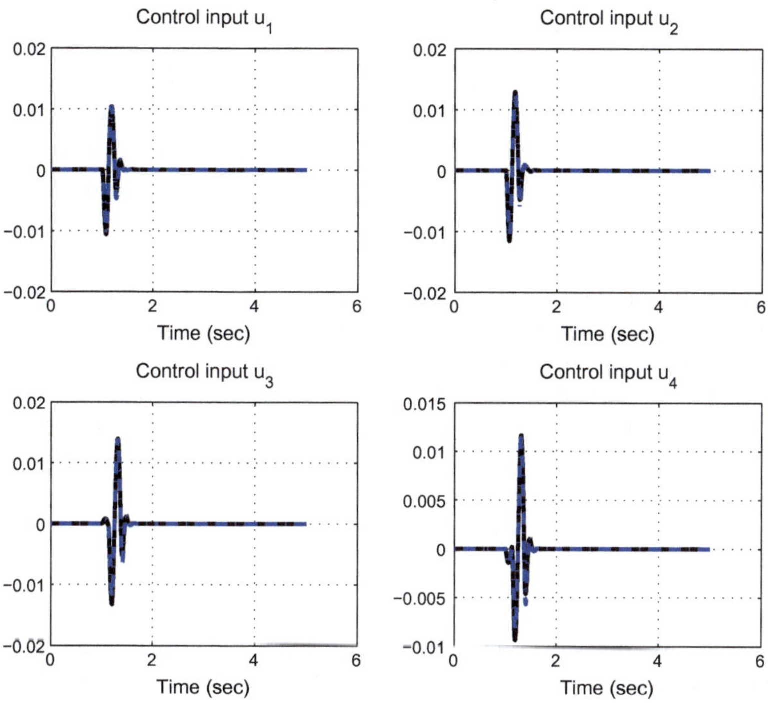

Fig. 7.12 Curves of the control inputs

Fig. 7.13 Responses of the displacements with different body masses ($M = 1200\,\text{kg}$, $1100\,\text{kg}$, $1300\,\text{kg}$)

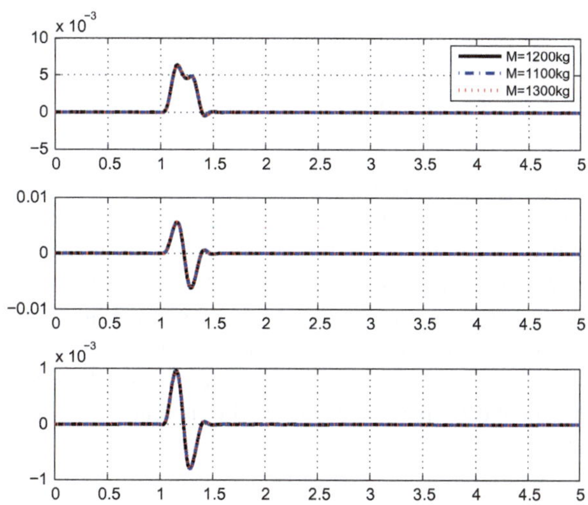

independent N(0, $2\pi n_0\sqrt{G_0 V}$) random variables, where G_0 stands for the road roughness coefficient and $n_0 = 0.1$ is the reference spatial frequency. Select the road roughness as $G_q(n_0) = 256 \times 10^{-6}\text{m}^3$, which is corresponded to D Grade (Poor)

Table 7.3 RMS values of body displacements

	Passive suspension	Active suspensions (B.C.)/improvements	Active suspensions (ARC)/improvements
y_{rms}	0.0037	0.0031(↓ 45.95%)	0.0020(↓ 56.52%)
θ_{rms}	4.6163×10^{-4}	2.2482×10^{-4}(↓ 51.30%)	2.0949×10^{-4}(↓ 54.62%)
ϕ_{rms}	4.1853×10^{-4}	1.7111×10^{-4}(↓ 59.12%)	1.0783×10^{-4}(↓ 74.24%)

according to ISO2631 standards, to generate the random road profile. For the random road disturbances, it is more reasonable to evaluate the effectiveness of controller design by using RMS values. It is well known that the RMS value of the vehicle body displacement or acceleration is strictly related to the ride comfort of passengers, and it is often used to quantify the amount of displacement or acceleration transmitted to the vehicle body. The RMS value of an n-dimensional vector x is calculated as:

$$x_{RMS} = \frac{\|x\|}{\sqrt{n}} = \sqrt{\frac{1}{n}\sum_{j=1}^{n} x_j^2}, \; j = 1, \ldots, n. \tag{7.47}$$

Table 7.3 gives the RMS values of the heave, pitch, roll displacements and accelerations, and from this table we can see that compared with backstepping-based H_∞ controllers, our proposed controllers can achieve a greater degree of reduction for the displacements and accelerations under the random road inputs, which further verifies the feasibility of proposed method.

7.2.4 Conclusion

This section focused on the research of the theory and methodology of ARC-based H_∞ control, and the problem of vibration suppression was investigated, where the electro-hydraulic actuators with highly nonlinear characteristics were considered for accurate control. An ARC-based H_∞ control was introduced to realize the nonlinear disturbance suppression and simultaneously increase the system robustness against both actuator parameter uncertainties and external disturbances. The stability analysis for the closed-loop system was given within the Lyapunov framework, and different road conditions were considered to reveal the performance of the controllers.

References

1. H. Merritt, *Hydraul. Control Syst.* (John Wiley & Sons Inc, New York, USA, 1967)
2. G. Matthijs, C. MarcvandeWal, B. Okko, LPV control for a wafer stage: beyond the theoretical solution. Control Eng. Pract. **13**(2), 231–245 (2005)

3. A. Alleyne, R. Liu, On the limitations of force tracking control for hydraulic servosystems. J. Dyn. Syst. Meas. Control **121**(2), 184–190 (1999)
4. S. Huang, W. Lin, Adaptive fuzzy controller with sliding surface for vehicle suspension control. IEEE Trans. Fuzzy Syst. **11**(4), 550–559 (2003)
5. B. Yao, M. Tomizuka, Adaptive robust control of MIMO nonlinear systems in semi-strict feedback forms. Automatica **37**, 1305–1321 (2001)
6. L. Lu, B. Yao, W. Lin, *A Two-loop contour tracking control for biaxial servo systems with constraints and uncertainties* (Proc. Amer. Control Conf, Washington, DC, 2013), pp. 6468–6473
7. G. Zhang, J. Chen, Z. Lee, Adaptive robust control for servo mechanisms with partially unknown states via dynamic surface control approach. IEEE Trans. Control Syst. Technol. **18**(3), 723–731 (2010)
8. F. Zhang, G. Duan, Robust adaptive integrated translation and rotation control of a rigid spacecraft with control saturation and actuator misalignment. ACTA Astronaut. **86**, 167–187 (2013)
9. A. Alleyne, J. Hedrick, Nonlinear adaptive control of active suspensions. IEEE Trans. Control Syst. Technol. **3**(1), 94–101 (1995)
10. N. Yagiz, Y. Hacioglu. Backstepping control of a vehicle with active suspensions. Control Eng. Pract
11. Z. Wang, Y. Liu, X. Liu, H_∞ filtering for uncertain stochastic time-delay systems with sector-bounded nonlinearities. Automatica **44**(5), 1268–1277 (2008)
12. S. Chantranuwathana, H. Peng, Adaptive robust force control for vehicle active suspensions. Int. J. Adapt. Control Signal Process. **18**(2), 83–102 (2004)
13. W. Sun, H. Gao, O. Kaynak, Finite frequency H_∞ control for vehicle active suspension systems. IEEE Trans. Control Syst. Technol. **19**(2), 416–422 (2011)
14. B. Yao, G. Chiu, J. Reedy. Nonlear adaptive robust control of one-dof electro-hydrulic servo systems. in *ASME International Mechanical Engineering Congress and Exposition (IMECE'SY)*
15. M. Ma, H. Chen, Disturbance attenuation control of active suspension with non-linear actuator dynamics. IET Control Theor. Appl. **5**(1), 112–122 (2011)
16. N. Yagiz, Y. Hacioglu, Backstepping control of a vehicle with active suspensions. Control Eng. Pract. **16**, 1457–1467 (2008)

Chapter 8
Energy Saving Control Strategies: Motor-Driven Active Suspension

One of the disadvantages of the active suspension is its high energy consumption, which limits its application. To overcome this problem, in recent years, researchers pay much attention to reducing energy consumption of active suspensions to make active suspensions widely used, and there are many remarkable results. A great deal of energy is dissipated in the form of heat through the viscous damper and the actuator. In other words, most of the vibration energy excited by road is transferred into useless heat energy. If we can recycle vibration energy excited by the uneven road in some way, it is attainable to consume less energy, even to achieve a self-powered suspension. So energy saving control strategies of active suspensions is a research field worth studying.

In this chapter, linear DC motors are chosen as the actuators to implement active control. In order to reduce energy consumption of active suspension, an energy regeneration scheme based on energy balance of the actuator is proposed. In the energy regeneration implementation scheme, both the operating zones of the actuator and working modes of the suspension are divided into three parts. In different zones or modes, energy consumption of the motor and the power source is various. By switching corresponding operating electric circuits, it is realizable to accumulate energy from road vibration and supply energy to the actuator. Moreover, the criterion of self-powered suspension is presented which can be employed to judge whether an active suspension can be self-powered or not. After the above analysis, a self-powered suspension with state feedback controller is designed. Simulation is conducted and results show that performances of the self-powered active suspension are nearly the same as those of the active suspension with external energy source, and this self-powered suspension can achieve energy regeneration with acceptant ride comfort.

W. Sun et al., *Advanced Control for Vehicle Active Suspension Systems*, Studies in Systems, Decision and Control 204,
https://doi.org/10.1007/978-3-030-15785-2_8

8.1 Analysis of Energy Flow

Analyzing the simplified armature circuit, the operating state of the motor can be divided into three zones. Based on the analysis of energy consumption and regeneration, we find the principle of energy-regenerative suspension. Further, the condition to achieve a self-powered suspension is obtained.

8.1.1 Energy Balance of DC Motor

A DC motor can work as an electromoter or generator. In active suspension control systems, allowing for the voltage of battery, the working states of the DC motor can be divided into 4 types [1], and relevant working circuits are as follows, where U_s is the supply voltage of the circuit, R is the resistance of the armature, E is induced voltage, i is armature current, and armature inductance is neglected.

(1) Electromotor state: the electric circuit of electromotor state is shown in Fig. 8.1. Supply voltage should be larger than induced voltage, and the direction of the two kinds of voltage should be opposite. The direction of armature current is opposite to that of induced voltage. However, electromagnetic force is exported in the same direction of motion of the motor, which means electromagnetic force contributes to the movement of the motor. In electromotor state, the motor converts electrical energy to mechanical energy, consuming energy from the power source.

(2) Generator state: the electric circuit of generator state is illustrated in Fig. 8.2. Here, supply voltage should be smaller than induced voltage, and the direction of

Fig. 8.1 Electromotor state

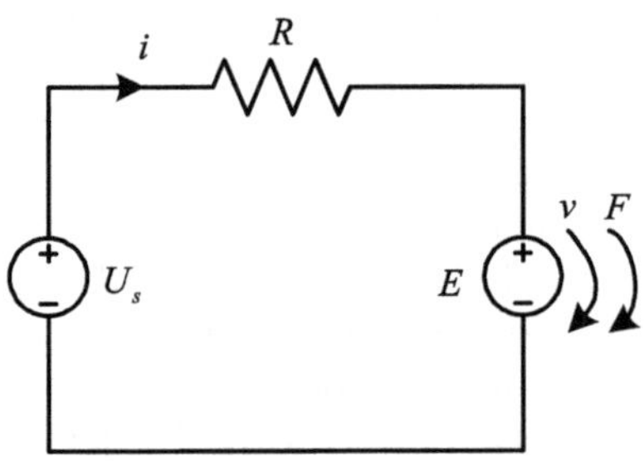

Fig. 8.2 Generator state

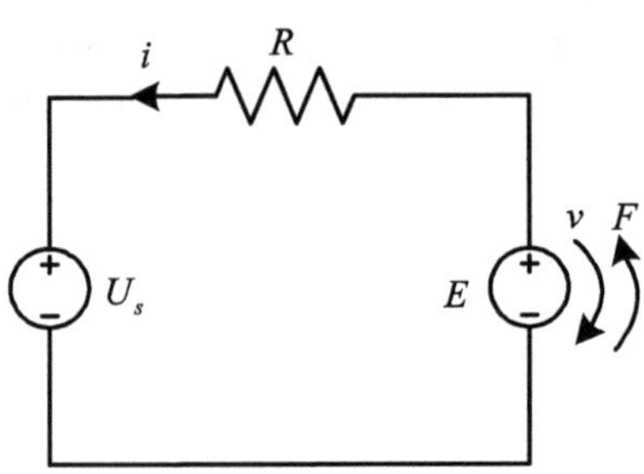

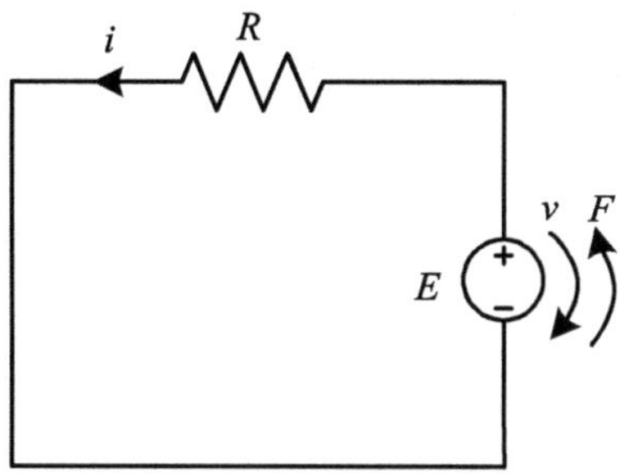

Fig. 8.3 Regenerative braking state

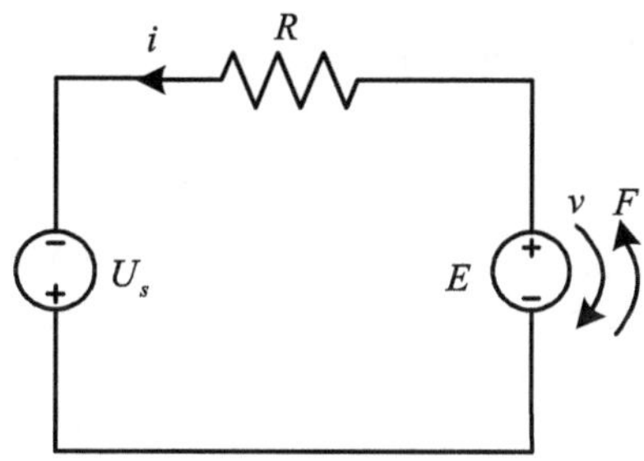

Fig. 8.4 Plug braking state

the two kinds of voltage should be opposite. The direction of armature current is the same as that of induced voltage. Electromagnetic force is exported in the opposite direction of motion of the motor, and that's because electromagnetic force opposes the motion of the motor. In generator state, the motor converts mechanical energy to electrical energy, charging the power source.

(3) Regenerative braking state: the electric circuit of regenerative braking state is presented in Fig. 8.3. Supply voltage is zero, and the motor is connected to short circuit. It can be seen as a particular case of generator state. The direction of armature current is the same as that of induced voltage. Electromagnetic force is exported in the opposite direction of motion of the motor, which makes the motor decelerate until the speed reaches zero.

(4) Plug braking state: the electric circuit of plug braking state is shown in Fig. 8.4. The direction of supply voltage and induced voltage should be the same. The direction of armature current is the same as that of induced voltage. Electromagnetic force is exported in the opposite direction of motion of the motor, and the braking effect is much more better than that in generator state or regenerative braking state. In plug braking state, the motor converts mechanical energy to electrical energy, and power source supplies energy to then motor in the meantime. All these energy including both mechanical energy and electrical energy is consumed by the armature resistant and transferred to heat loss. Due to the high current and large output force, some measures should be used to avoid the over current.

The simplified armature circuit of the motor [2] is shown in Fig. 8.5. In this figure, the voltage balance equation of the circuit can be obtained as

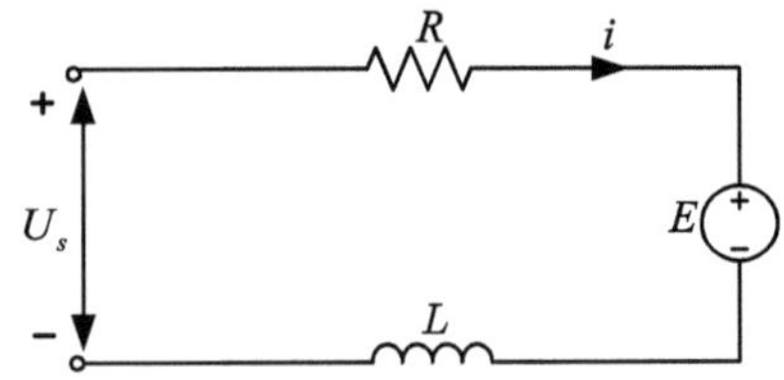

Fig. 8.5 The simplified armature circuit of the motor

$$U_s = R \cdot i + L \cdot \frac{di}{dt} + E \tag{8.1}$$

Assuming induced voltage is negative when the rotor moves upward, then we can get:

$$E = -k_E \cdot v \tag{8.2}$$

where K_e represents the back EMF coefficient and v is the relative speed of the suspension. Regardless of the effect of the inductance, then, combing (8.2), (8.1) can be rewritten as:

$$U_s = R \cdot i + k_E \cdot v \tag{8.3}$$

Let P_a, P_b be the power of the motor and the power consumed by the power supply respectively, and they can be described by the following equations:

$$P_a = -E \cdot i = k_E \cdot v \cdot i \tag{8.4}$$

$$P_s = U_s \cdot i \tag{8.5}$$

$P_a > 0$ indicates that the motor works as a electromotor and electrical energy is converted to mechanical energy, while $P_a < 0$ indicates that the motor works in a generator operating condition and mechanical energy is converted to electrical energy. Similarly, the power supply consumes energy when $P_s > 0$ and $P_s < 0$ implies that the power source is being charged [3].

The motor produced control force F:

$$F = k_i \cdot i = k_i \frac{E}{R} \tag{8.6}$$

When adopting the international system of units, the following equation describes the relationship between the back EMF coefficient and motor torque coefficient:

$$k_E = k_i \tag{8.7}$$

Based on (8.2) and (8.6), we obtain

$$F = -k_i{}^2 \frac{v}{R} \tag{8.8}$$

The expression of control force has the same form as the one of viscous damping force, and c_{eq} is defined as equivalent damping coefficient [4]:

$$c_{eq} = -\frac{{k_i}^2}{R} \tag{8.9}$$

To some extent, the motor can be viewed as a kind of viscous damper when the connection from the power supply is cut off and the two ends of the motor is directly connected. That case means the active suspension with a DC motor can work as a passive suspension without external voltage input.

8.1.2 Operating Zones of Motor

According to the working status of the motor and power source, the operating state of the motor can be divided into 3 zones [2]. The mathematical expression of the straight line in Fig. 8.6 is given:

$$i = -\frac{k_E}{R}v \tag{8.10}$$

When control current is above the straight line, we can get $U_S = R \cdot i + k_E \cdot v > 0$. Conversely, i below the straight line means $U_S < 0$.

When the motor operates in Zone I ($P_a > 0,\ P_s > 0$) as shown in Fig. 8.6, the motor works as an electromotor which consumes energy supplied by the power source. In Zone II, the motor is in the generator state while the power source still consumes energy. In Zone III, the motor regenerates mechanical energy excited by uneven road and the power source is charged by the motor in the meanwhile. Therefore, we can see that only in Zone III can energy be regenerated, and meanwhile can this part of energy be stored and utilized afterwards so that the goal of saving energy can be achieved.

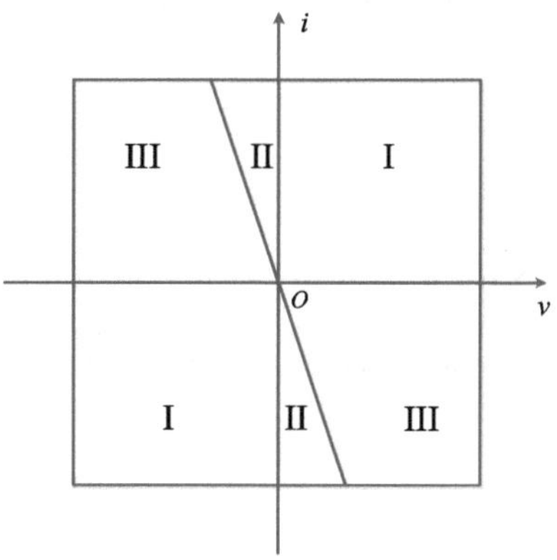

Fig. 8.6 The operating zones of the motor

Energy regenerative active suspension can be seen as a special kind of active suspension in consideration of motors' operating state. Electric circuit for energy regeneration needs to be added to active suspension to store and supply electrical energy harvested by the motor.

8.2 Criterion of Self-powered Suspension

If the motor locates in Zone I or Zone II, the suspension consumes energy. If the motor locates in Zone III, the motor outputs energy to the power source. On the occasion of road disturbance, the energy consumption and output decide the potential of the active suspension to regenerate energy. The more energy regeneration and the less energy consumption, the better energy efficiency can be realized.

Supposing energy output is more than energy consumption, replace the power source with a condenser, and we get a self-powered suspension whose performance will be close to that of the active suspension without energy regeneration capacity. Different motor parameters and controller gains can result in different energy conversion status. Through choosing proper motor and controller, self-powered suspension is possible to be obtained.

The average energy consumption of the power source can be expressed as follows [4]:

$$\bar{E}_s = \frac{1}{\pi}\int_0^{\infty} e_s(\omega)G_0(\omega)d\omega \tag{8.11}$$

where $G_0(\omega)$ denotes the power spectra density of road disturbance. $e_s(\omega)$ stands for power transfer function of the power source, which is defined as:

$$e_s(\omega) = R|G_i(j\omega)|^2 + k_i\,|G_v(j\omega)|\,|G_i(j\omega)|\cos(\phi_v(\omega) - \phi_i(\omega)) \tag{8.12}$$

where $G_i(j\omega)$ and $G_v(j\omega)$ are the frequency responses of control current i and suspension relative speed v respectively. And $\phi_i(\omega)$, $\phi_v(\omega)$ represent the phase angle of i and v respectively. If $G_0(\omega) = 1$, the disturbance input is an ideal white noise whose intensity equals 1, and (8.11) can be simplified as:

$$\bar{E}_s = \frac{1}{\pi}\int_0^{\infty} e_s(\omega)d\omega \tag{8.13}$$

(8.13) is defined as the criterion to attain self-powered suspension: if $\bar{E}_s < 0$, the active suspension can be self-powered through regenerative energy without the external power source, whereas the active suspension cannot be self-powered in the case of $\bar{E}_s > 0$.

8.3 Energy Regeneration Implementation Scheme

In order to realize energy regeneration of a self-powered suspension, relevant strategies should be taken to deal with different working states of the DC motor. Based on analyzing energy flow in the electric circuit, a mode variable [4] is found to identify the operating zone which the motor works in. On basis of the mode variable, three working modes of the suspension are given, and the operating circuit for each working mode is designed separately. An integrated energy regeneration implementation scheme is developed systematically in this section.

The mode variable [4] which can be utilized to identify the operating zone of the motor is defined as follows:

$$\gamma = \frac{R}{k_i} \cdot \frac{i}{v} (v \neq 0) \tag{8.14}$$

Applying the expression of γ to (8.4), the power of the motor can be depicted as:

$$P_a = k_i \cdot v \cdot i = \frac{k_i^2}{R} \cdot \frac{R}{k_i} \cdot \frac{i}{v} \cdot v^2 = c_{eq} \gamma v^2 \tag{8.15}$$

Similarly, the power consumed by the power supply becomes

$$P_s = U_s \cdot i = (Ri + k_i v)i = c_{eq} \gamma (\gamma + 1) v^2 \tag{8.16}$$

Different value ranges of γ correspond to three operating zones of the motor, and three working modes of the energy-regenerative active suspension: drive mode, brake mode, and regeneration mode, as concluded in the following Table 8.1.

The operating electric circuit of the energy-regenerative suspension is shown in Fig. 8.7. In Fig. 8.7, R_{var} represents variable resistor, and its value is controlled by computer. Through changing the value of the variable resistor, control current can be regulated to track the desired current i^* computed by controller. The condenser C is adopted to store and supply energy, taking place of the power source in active suspension. Switch S is used to turn on or turn off the circuit. Two groups of controllable switches symbolized by A and B are utilized to switch among the three suspension working modes. The motor can be connected to the condenser or be shorted by switches in group A ($S_{A1} - S_{A3}$). The role of switches in group B ($S_{B1} - S_{B4}$) is to change the direction of control current to match the operating zones of the motor.

Table 8.1 Relationships among γ, Operating Zones of the Motor, and Working Modes of the Suspension

The range of γ	Operating zone	Power	Working mode
$\gamma \geq 0$	I	$P_a \geq 0$, $P_s \geq 0$	Drive mode
$\gamma \leq -1$	II	$P_a < 0$, $P_s \geq 0$	Brake mode
$-1 < \gamma < 0$	II	$P_a < 0$, $P_s < 0$	Regeneration mode

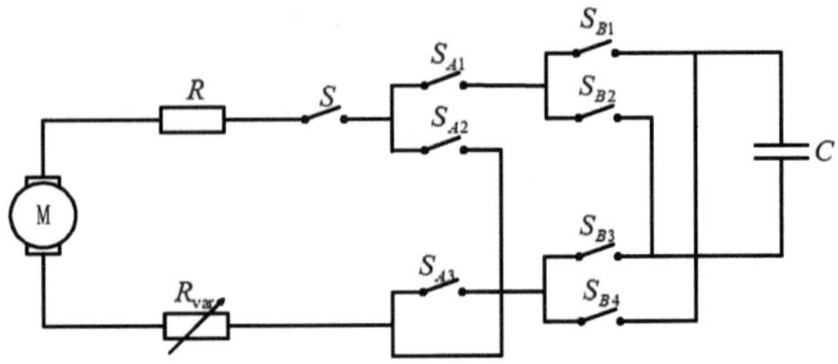

Fig. 8.7 The operating electric circuit of the energy-regenerative suspension

The value of γ can be calculated through state variables detected by sensors and desired control current computed by controller. In this way, we can judge the working mode of the suspension, and switch to corresponding control circuit so that self-powered suspension is realizable.

Mode I: Drive Mode

When $\gamma \geq 0$, the motor and the suspension work in Zone I and Mode I respectively. Energy stored in the condenser is utilized to drive the motor to provide control force in the reverse direction of the damping force. Control current can be figured by (8.17)

$$i = \frac{\sigma e_c - k_i v}{(R + R_{var})} \tag{8.17}$$

where e_c represents the voltage of the energy-storage condenser, and a scalar σ is defined as a symbol variable, indicating the sign of e_c:

$$\sigma = \begin{cases} -1 & v > 0 \\ 1 & v < 0 \end{cases} \tag{8.18}$$

Through modifying R_{var}, the desired control current can be tracked. So we can calculate the resistance of the variable resistor from desired control current i^*, as (8.19) follows:

$$R_{var} = \frac{1}{i^*}(\sigma e_c - k_i v) - R \tag{8.19}$$

where i^* is a variable with a positive or negative sign.

On the occasion of $e_c > k_i |v|$ and $R_{var} > 0$, if $v > 0$, close the following controllable switches: S_{A1}, S_{A3}, S_{B1}, S_{B3}, and if $v < 0$, close S_{B2}, S_{B4} instead and the switching operation case of group A remains unchanged. For convenience, we dont mention the direction of suspension relative speed later in the paper. When R_{var} is minus, which means the voltage of the condenser is not high enough to provide the desired control current, the value of the variable resistor is set to zero. In this case, the actuator cant provide desired force. If $e_c < k_i |v|$, open switch S to cut the circuit.

Mode II: Brake Mode

When $\gamma \leq -1$, the motor works in Zone II and we can determine that the suspension operates in Mode II. The actuator produced control force with energy generated by its own and energy stored in the energy-storage condenser. Then, the control current can be expressed as:

$$i = \frac{-\sigma e_c - k_i v}{R + R_{var}} \tag{8.20}$$

R_{var} can be obtained from desired control current:

$$R_{var} = \frac{1}{i*}(-\sigma e_c - k_i v) - R \tag{8.21}$$

When $e_c \geq 0$, controllable switches S_{A1}, S_{A3}, S_{B2}, S_{B4} are closed in Fig. 8.7. Set the value of R_{var} to zero once it becomes negative. When $e_c < 0$, in order to avoid applying reverse voltage to the condenser, cut the connection between the actuator and the condenser, connect the motor to the short circuit, and let R_{var} be zero at the same time.

Mode III: Regeneration Mode

In the case of $-1 < \gamma < 0$, the motor and the suspension work in Zone III and regeneration mode respectively. One part of energy converted from road vibration is utilized to produce damping force, and the other part is accumulated to charge the condenser. The control current is decided as:

$$i = \frac{\sigma e_c - k_i v}{R + R_{var}} \tag{8.22}$$

The resistance of the variable resistor is written as:

$$R_{var} = \frac{1}{i*}(\sigma e_c - k_i v) - R \tag{8.23}$$

R_{var} is assigned to zero, and the operating electric circuit is the same as the one in drive mode when $e_c > k_i\,|v|$. When R_{var} is calculated to be negative by (8.23), short the motor and the motor plays a role as a viscous damper. Then R_{var} becomes:

$$R_{var} = \frac{-k_i v}{i*} - R \tag{8.24}$$

In this situation, only switch S and switch S_{A2} are closed.

8.4 Simulation Verification

The suspension is a complicated vibration system. For the convenience purpose, a two-degree-of–freedom suspension: a quarter-car model is applied here. The state equations of the quarter-car model are given by (2.8), and matrixes are defined as (2.2) and (2.9) in Chap. 2.

We adopt the parameters of the suspension system listed in Sect. 2.1 where a entire frequency constrained H_∞ scheme is suggested to be used. In Sect. 2.1, the state feedback controller is designed as (2.44):

$$K = 10^4 \times [\,1.0098\ 4.9655\ -0.1896\ 0.0909\,]$$

According to (8.13), we can calculate the average energy consumption of the power source, that is:

$$\bar{E}_s = -2.7642 \times 10^5 < 0 \tag{8.25}$$

which meets the condition of self-powered suspension. Therefore, it is possible for the active suspension to be self-powered. The energy regeneration implementation scheme applied to the suspension is the one stated above. A self-powered active suspension is got.

Simulations are conducted under random road excitation in the form of white noise. The power spectra density of uneven road velocity input is described as (2.9), that is :

$$G_{\dot{q}}(f) = 4\pi^2 G_q(n_0) v n_0^2$$

Also, let the road class be C, with road roughness coefficient of $256 \times 10^{-6}\,\mathrm{m}^3$, and the vehicle drives at the speed of 45 km/h.

The simulation results of the self-powered active suspension are compared with those of ordinary active suspension. From Fig. 8.8, the vehicle heave acceleration of

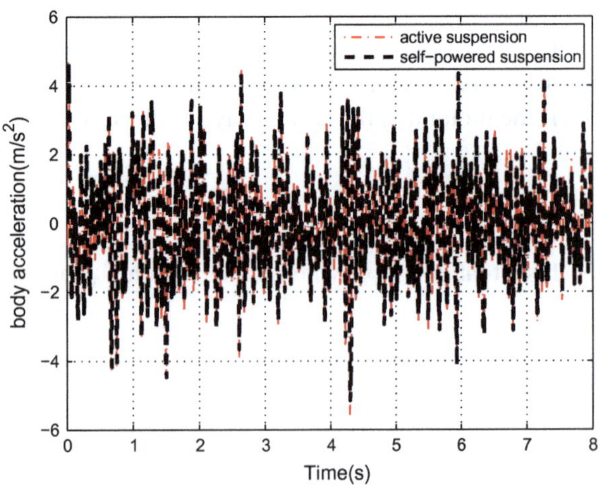

Fig. 8.8 The vehicle heave acceleration of self-powered suspension compared with active suspension

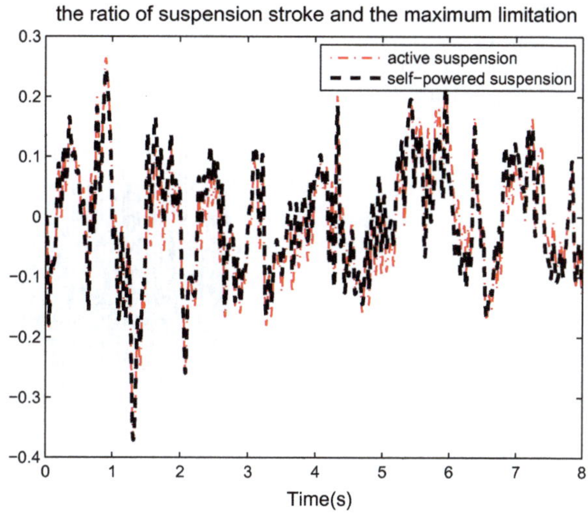

Fig. 8.9 The suspension stroke of self-powered suspension compared with active suspension

the self-powered suspension is close to that of the active suspension, whereas body accelerations of the self-powered suspension are a little bigger at certain moments. It can be indicated that both self-powered suspension and active suspension can provide good ride comfort, even though some ride comfort is sacrificed on account of energy regeneration. Curves of suspension strokes about the two kinds of suspensions are generally similar, which is described in Fig. 8.9. In terms of dynamic tire deflection, time domain responses of the self-powered suspension and active suspension without energy regeneration are almost the same, as in Fig. 8.10. The ratios of these two constraints and their limits are both below 1, with regards to the self-powered active suspension and active suspension. These results exactly indicate performances of self-powered suspension are nearly consistent with those of active suspension.

Control force produced by the actuator without energy control and the desired force are of equal value. However, for self-powered suspension, the actual force may not always equal the value of desired force. From Fig. 8.11, we can see active control force of the self-powered suspension track the desired control force roughly. Only on several points in Fig. 8.11, are values of actual force slightly smaller than those of desired force. Admittedly, there is a trade-off between ride comfort and energy regeneration.

It is assumed that the vehicle is driving for 8 seconds, and energy consumed by the condenser in this period is

$$W_C = \int_0^8 e_c \cdot idt = -305.7335J < 0 \tag{8.26}$$

Compared with ordinary active control, we can conclude that self-powered suspension provides a good capacity of energy regeneration with acceptable ride comfort.

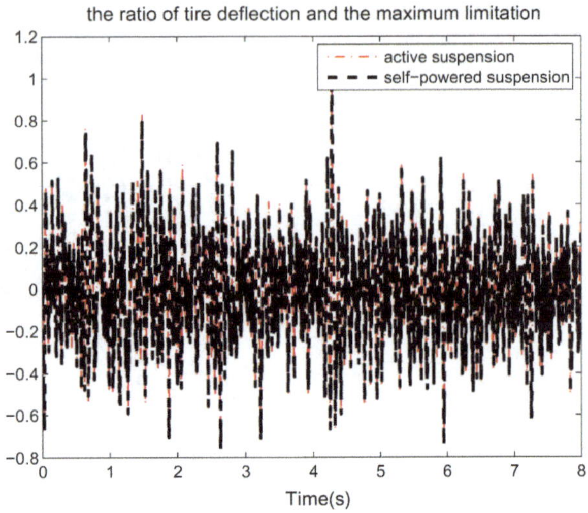

Fig. 8.10 The dynamic tire deflection of self-powered suspension compared with active suspension

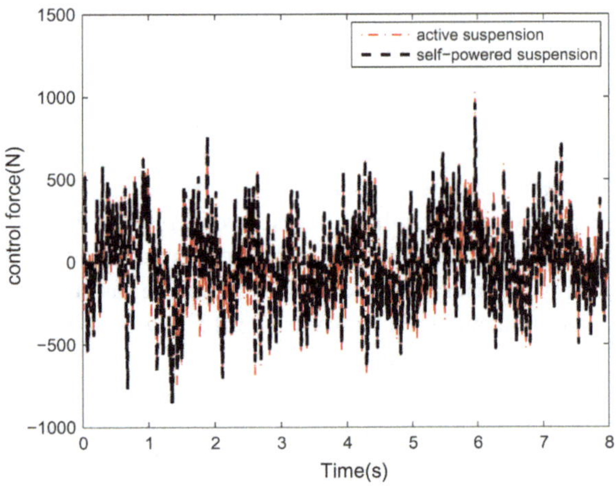

Fig. 8.11 The active control force of self-powered suspension compared with active suspension

8.5 Conclusion

In this chapter, the study is focused on energy regeneration of the motor-driven suspension. A linear DC motor is selected to be the actuator of the suspension. Based on analyzing energy flow in the motor executive circuit, the operating state of the motor is divided into 3 zones. In different zones, energy consumption of the motor and the power source is in different situations. The criterion to realize a self-powered suspension is proposed, and whether an active suspension can be self-powered or not will be determined by calculating the average energy consumption of the power source. To reclaim and supply energy, an energy regeneration implementation scheme is demonstrated to make sure the related physical process is available, where the working mode of the suspension is identified through the mode variable and controllable switches are used to connect different operating electric circuits.

A two-degree-of-freedom suspension is chosen as the model to study. In the design example, the selection of the suspension system parameters and design of the controller meets the prerequisite for self-powered suspension. From simulation results, it can be concluded that performances of the self-powered active suspension are nearly the same as those of the active suspension with an external energy source. The effectiveness of self-powered suspension is proved. The self-powered suspension is effective in energy regeneration with maintaining good ride comfort in the meantime.

References

1. X. Mei, G. Bai, M. Zhang, *Automatic control element and circuit.* (Science Press, Beijing, 2007)
2. G. Zhang, J. Cao, Y. Fan, Design of active and energy-regenerative controllers for DC-motor-based suspension. Mechatronics **22**, 1124–1134 (2012)
3. K. Huang, F. Yu, Y. Zhang, Active controllable design for an electromagnetic energy-regenerative suspension. Int. J. Autom. Technol. **12**(6), 877–885 (2011)
4. K. Nakano, Y. Suda, S. Nakadai, Self-powered active vibration control using a single electric actuator. J. Sound Vib. **260**, 213–235 (2003)

MIX
Papier aus verantwortungsvollen Quellen
Paper from responsible sources
FSC® C105338

If you have any concerns about our products,
you can contact us on
ProductSafety@springernature.com

In case Publisher is established outside the EU,
the EU authorized representative is:
Springer Nature Customer Service Center GmbH
Europaplatz 3, 69115 Heidelberg, Germany

Printed by Libri Plureos GmbH
in Hamburg, Germany